Professor Bear's Math Club
Coach Manual Alpha

AF882385

Marjorie Sayer

Atelier Finwhale

ISBN: 0988256118
ISBN-13: 978-0-9882561-1-8

Preface

This series of books arose from my work as a volunteer math coach for an elementary school math club.

The emphasis of this book is **not** the elementary school math curriculum;
not "bringing kids up to international standards;"
not "something relevant to the 21st century."

The emphasis **is**: engaging in mathematics as a life-affirming creative pursuit.

This is a short way of saying that mathematics is fun; it connects you to other people who are interested in having fun; it gives you opportunities to use the math parts of your brain—and everyone has them.

When you do fun math, when you connect with other people doing fun math, and when you use your math abilities to your full potential, mathematics makes you feel glad to be you.

It is my great pleasure to write this book with no other ideal in mind.

RRRRR.

Professor L. F. Bear
Palo Alto, 2012

Acknowledgements

Professor L. F. Bear gratefully acknowledges the inspiration provided to her by the members of the El Carmelo Math Club: Arjun, Ava, Chris, Ciel, Clara, Harrison, Jeffrey, Jonathon, Kristina, Lauren, Lia, Miranda, Moeka, Noam, Sarah, Sophia.

For encouragement and valuable insights, Professor Bear owes much to Jeannette Cheng, Gabrielle Conway, Betsy Franco, Alan Liu, Heidi Mickelsen, Ken Tracton, Maya Venkatraman, and Stephanie Youngquist. Thank you so very much.

Any errors or omissions are due to Marjorie Sayer.

How To Use This Manual

This manual is a companion book for *Professor Bear's Math Club Workbook Alpha.* It contains complete solutions, alternate solutions, and background information on the topics covered in *Workbook Alpha*.

The manual has ten chapters, following *Workbook Alpha*. Every chapter has this structure:

Title	Tells you what the topic is.
Why <Topic>	Professor Bear's reasons for including this topic in the book, in the order it is presented. This section also describes the topic in the larger context of mathematics, and what further topics are based on this topic.
Underpinnings Of <Topic>	The math you need to know in order to successfully navigate this chapter. It's a way to determine if a particular person is ready for the material in the chapter.
Outcomes	The mathematical goals and take-aways of this chapter.
Solutions: Notes and Vocabulary	Solutions of the *Notes And Vocabulary* sections of *Workbook Alpha*.
Solutions: Problems	Solutions of the ten problems in *Workbook Alpha*.

See also: http://bear.mathbelt.com

Contents

1. Sequences

Why Sequences

Here are Professor Bear's reasons to begin with a chapter about Sequences:

- Sequences are ordered lists that might have patterns. They are a great way to study patterns and only require a little knowledge of numbers and arithmetic to get started.
- Sequences are elementary functions. They map counting numbers 1, 2, 3, ... to other numbers or items.
- Sequences are usually not given top billing in math class, yet they often appear on standardized tests.
- Sequences have a long and rich mathematical history. If you become a math major you will come back to them, and wonder why you didn't spend more time on them earlier. They are accessible to very young mathematicians.
- Monthly bank statements are sequences. Mortgage payments are sequences. Mortgage payments are calculated from geometric sequences. Drug treatments are sequences. Musical transcriptions are sequences. Sequences are pervasive.

Underpinnings of Sequences

To understand this material you need some experience with arithmetic and be able to do one and two digit addition, subtraction, multiplication and division.

It's also helpful but not necessary to know fractions.

Outcomes

After doing this chapter the student will have more experience with sequences and finding number patterns. Students might find themselves motivated to learn basic number facts and fractions well enough to be able to play with sequences.

Solutions: Notes and Vocabulary

A **sequence** is an ordered list of things.

Here is a sequence of numbers: 1, 2, 3, 4, 5
Here is a sequence of letters: A, B, C, D
Here is a different sequence of numbers: 5, 4, 3, 2, 1

The "things" in the sequence are called **terms**. In the sequence 5, 4, 3, 2, 1
- first term is 5
- the third term is 3
- the fifth term is 1.

What is the first term of the sequence 1, 2, 3, 4, 5? **1**

What is the fourth term of the sequence A, B, C, D? **D**

A sequence can be finite or infinite. The sequences above are all finite.

Here's another example of a finite sequence: 10, 9, 8

Here's another example: 400, 500, 600, 700, 800

Write down your own example of a finite sequence: ____________________
Have club members check one another's examples.

An infinite sequence is a list that goes on forever. It has an infinite number of terms. We show that it goes on forever by three periods in a row at the end: ...

Another name for "three periods in a row" is ellipsis.

Here is an example of an infinite sequence: 1, 2, 3, ...

Another example of an infinite sequence: $\frac{1}{2}, \frac{2}{3}, \frac{3}{4} \ldots$

Write down your own example of an infinite sequence: ____________________
Have club members check one another's examples.

A sequence of numbers might follow a pattern. If you know the pattern, you can figure out what the next term is. Can you write the next few terms of these sequences?

2, 4, 6, 8, 10, 12 , ...

4, 7, 10, 13, 16, 19 , ...

Some sequence patterns might be more challenging, such as this finite sequence:

3, 2, 1, 2, 3, 3, 3, 2, 2, 2, 3, 5, 5, ... , 1

Can you fill in the missing terms? Think for a minute. A hint for this sequence is given later.

You can also use an ellipsis to write down a missing set of terms in a finite sequence.
For example: 1, 2, 3, ... , 10
Is a short way to write 1, 2, 3, 4, 5, 6, 7, 8, 9, 10
In the example 1, 2, 3, ... , 10 the ellipsis stands for the missing terms 4, 5, 6, 7, 8, 9

In the following finite sequences, what are the missing terms?

Sequence	Missing Terms
10, 9, 8, ... , 4	**7, 6, 5**
0, 3, 6, ... , 24	**9, 12, 15, 18, 21**
77, 72, 67, ... , 47	**62, 57, 52**

Hint for the sequence 3, 2, 1, 2, 3, 3, 3, 2, 2, 2, 3, 5, 5, ... , 1 above: musical notes. The sequence corresponds to a melody. If you don't have musical experience, this example will not mean much to you.

The numbers correspond to the tune *Mary Had a Little Lamb.*

In that case, try to figure out the following pattern:

0, 3, 8, 15, 24, 35, 48, . . .

The pattern is to add progressively larger odd numbers.
0, 0+3, 3+5, 8+7, 15+9, and so on.

Solutions: Problems

1-1. Sample Sequences
Here are some sequences. What are the next three terms?

a. 5, 5.3, 5.6, 5.9, 6.2, . . . **Solution: 6.5, 6.8, 7.1**

b. 2, 6, 18, 54, 162, . . . **Solution: 486, 1458, 4374**

c. 2, 5, 9, 14, 20, . . . **Solution: 27, 35, 44**

d. one, one thousand, one million, . . . **Solution: one billion, one trillion, one quadrillion**

1-2. Consecutive Terms
Consecutive terms are terms that exactly follow one another in a sequence.
In the sequence 1, 2, 3, ... the terms 3, 4, 5, 6, and 7 are consecutive terms.
In the sequence A, B, C, ... , Z the terms P and Q are consecutive terms.

Note to Coach: Example solutions are entered below, but of course many valid solutions are possible. Now it is a good time to have students check one another's solutions—see if they are truly consecutive—and review the terminology. This problem is useful for group discussion.

Write three consecutive whole numbers: **1, 2, 3 or 56, 57, 58 or 999, 1000, 1001**

Write three consecutive whole numbers bigger than a million:

1000001, 1000002, 1000003 or 2000000, 2000001, 2000002

Write three consecutive odd numbers: **13, 15, 17 or 91, 93, 95**

Write three consecutive multiples of five: **5, 10, 15**

Write three consecutive prime numbers: **23, 29, 31**

Write three consecutive two-digit primes: **37, 41, 43**

Write three consecutive perfect squares: **4, 9, 16** or **49, 64, 81**

Pose a more challenging problem involving consecutive numbers that you are pretty sure you are able to solve. Make it as different from the above problems as you can.

Example: Write three consecutive four-digit multiples of 17.

Pose an even more challenging problem involving consecutive numbers that you are pretty sure you are not able to solve (yet).

Example: What are the dates of the next three consecutive supernova explosions that will happen in the Milky Way?

1-3. Fibonacci Sequence

The Fibonacci sequence is: 1, 1, 2, 3, 5, 8, 13, . . .

a. The terms of the Fibonacci sequence follow a pattern. Describe the pattern.

Solution: After the first two terms, each term is the sum of the previous two terms.

b. List the next three terms of the Fibonacci sequence.

Solution: 21, 34, 55

c. The terms of the Fibonacci sequence are called Fibonacci numbers. How many prime Fibonacci numbers can you find?

Solution: Fibonacci primes under 100 are: 2, 3, 5, 13, 89. The next one is 233.

d. Choose any three consecutive terms of the Fibonacci sequence. Square the middle term and multiply the two other terms. Compare the results. Repeat this process with other groups of three consecutive terms. What is the pattern? (This is Cassini's Identity.)

Solution: For any three consecutive terms, the square of the middle term has a difference of one from the product of the two outer terms.

1-4. Zeno's Paradox In A Box

The sequence $\frac{1}{2}, \frac{1}{4}, \frac{1}{8}, \ldots$

Is illustrated in this box:

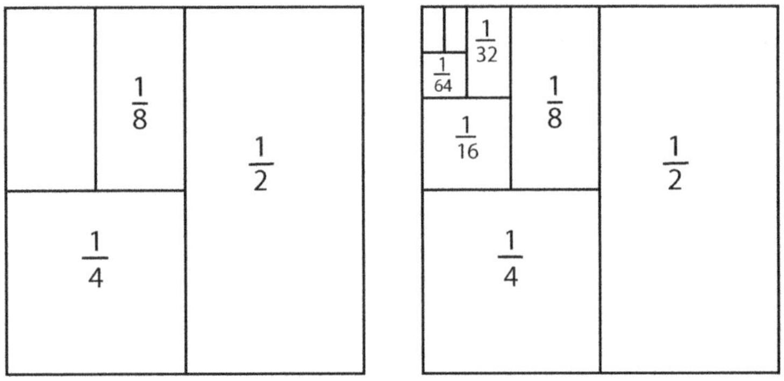

Draw in a few more terms in the box. **(Solution above.)**

Describe what happens to the terms of the sequence, as they go on and on.
The terms get smaller and smaller.

Here is another sequence: $\frac{1}{2}, \frac{1}{2}+\frac{1}{4}, \frac{1}{2}+\frac{1}{4}+\frac{1}{8}, \ldots$

Illustrate this sequence in these boxes:

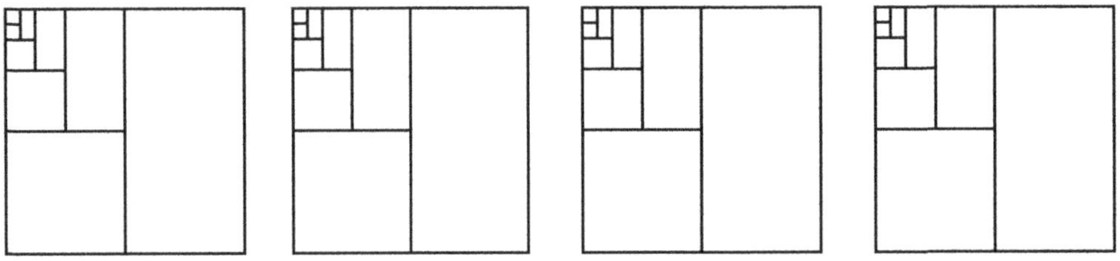

Solution:

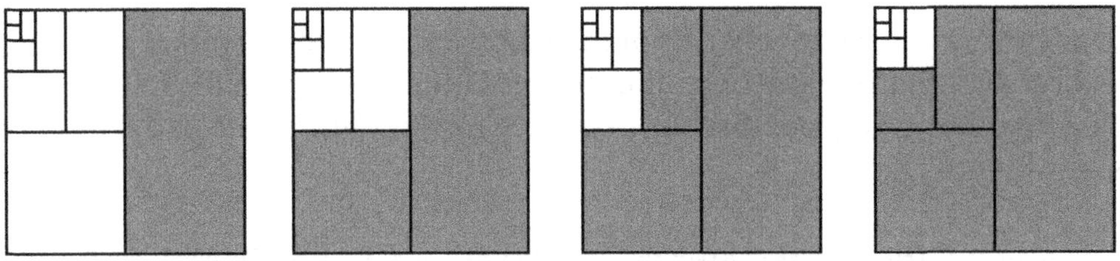

Describe what happens to the terms of this sequence, as they go on and on.
If you know how to add fractions, work out the terms and see if they follow a pattern.

Solution: The pattern is: ½, ¾, 7/8, 15/16, …
The terms of the sequence get closer and closer to 1. They fill the box.

1-5. Days Of The Week

The days of the week form a sequence that repeats every seven terms.
a. What day of the week is today? ______________ **(variable answers)**

b. What day of the week will it be 100 days from now?

Solution: Because 98 is a multiple of 7, the day of the week will be the day 2 days later from the current day. For example, if today is Monday, it will be Wednesday 100 days from now.

c. 500 days from now, will it be a different day of the week from today?

Solution: 500 days from now, the day of the week will be three days later than the current day, because 497 is a multiple of seven.

d. Name a three-digit number for which you know that it will be the same day as today, that many days from now.

Solution: 777 or any 3-digit multiple of 7. There are several ways for people to understand this problem. One way is about remainders when you divide by 7. If the current day is Monday, then the 8th day is Monday, and the 15th, and so on—any number with a remainder of 1 when you divide by 7. Or, forgetting about remainders, simply think of multiples of 7. In this problem, it is good to make sure everyone understands what is meant by "that many days from now." Sometimes people get confused: if it is Monday, is the next Monday 8 days away or 7 days away? (It becomes Monday again after 7 full days have passed.)

Another good strategy is to list the first several days and the corresponding "number of days away." For example, if today is Monday, then Tuesday is 1 day away, Wednesday is 2 days away, and so on.

1-6. Counting Terms

This problem is about careful reading of what the question is asking, and careful reading of the sequence pattern.

a. How many terms are in the sequence: 1, 3, 5, 7, . . . , 993, 995, 997, 999 ?

Solution: One approach (chunking): count the number of odd numbers less than 100, and multiply that result by 10, because the problem is asking for the number of odd numbers less than 1000. The number of odd numbers less than 10 is 5, so the number odd numbers less than 100 is 50. Ten times that result is 500.

Table approach: Make a table that starts with a counter. (If your students have access to a computer, it is fun to try to make this kind of table with Microsoft Excel.)

Counter	Convert the Counter	Odd number
1	1	1
2	1+2	3
3	1+2+2	5
4	1+2+2+2	7
5	1+2+2+2+2	9

With a table like the one above, students might be able to see a pattern. The third odd number is 1 + two twos. The fifth odd number is 1 + four twos. Continuing the pattern, the 100th odd number is 1 + ninety-nine twos.

Knowing the pattern, it is possible to work backwards from an odd number to the counter. In words, the pattern is:

Nth odd number = 1 + (N-1) twos

For example, if the odd number is 57, what is N? Working backwards from 57 = 1 + (N – 1) twos:

57 – 1 = 56 = (N – 1) twos

56 = 28 twos. So N = 29.
57 is the 29th odd number.

In the same way, 999 – 1 = 998.
998 = 499 twos. So N = 500.

Note to Coach: Some students are able to carry the many steps of this problem through to the end, and some get value out of working out the meaning of the 10th odd number or 20th odd number. It's very good to solve a simpler problem in the course of attempting this one.

b. Find the 1000th term in the sequence: $2, 3, 4, 5, 6, \ldots$

Solution: This sequence counts up from 2. The first term is 2, the second term is 3, the third term is 4. Following that pattern, the 1000th term is 1001.

c. Find the 1000th term in the sequence: $2, 4, 6, 8, \ldots$

Solution: This sequence is the sequence of even numbers. Using the chunking method: the number of even numbers up to 10 is 5; the number of even numbers up to 100 is 50. Therefore there are 100 even numbers up to 200. The 1000th even number is 2000.

Using the counter method:

Counter	Convert the Counter	Even Number
1	1x2	2
2	2x2	4
3	3x2	6
4	4x2	8

The pattern is that the 3rd even number is 3x2, the 4th even number is 4x2, and so on. So the 1000th even number is 1000x2 = 2000.

1-7. Cumulative Peanuts

Ella the elephant ate 1000 peanuts from Monday through Friday. Each day she ate 10 peanuts more than the previous day. How many peanuts did she eat on Friday?

Solution: Method 1: Write down how many peanuts Ella ate each day: Monday, X peanuts. Tuesday, X +10. Wednesday: X + 20. Thursday: X + 30. Friday: X + 40. Adding them all up, you get 5X + 100. So 5X is 900 and Ella ate 180 peanuts on Monday, which means she ate 220 peanuts on Friday.

Method 2: If she ate 1000 peanuts in 5 days, on average she ate 200 peanuts per day. Since there is an odd number of days, she eats the average on the middle day, Wednesday (before Wednesday, she eats less than average; after Wednesday, she eats more than average). If she eats 200 on Wednesday, she eats 220 on Friday.

1-8. Amoebas

An amoeba reproduces by dividing into two. Suppose it takes a day for an amoeba to eat enough food to be able to divide, and a scientist drops an amoeba into a pond in his back yard on January 1.

a. Write down the sequence of amoeba population values for January 1, January 2, and so on.

Solution: On January 1, the number of amoebas is 1. Each day thereafter they double. The sequence is 1, 2, 4, 8, 16, …

b. The scientist notices that the pond is full of amoebas on February 10. On what date would the pond be full if he dropped two amoebas instead of one on January 1?

Solution: There is a tendency to think that it would take half the amount of time to reach the same number of amoebas. But the sequence of populations looks like: 2, 4, 6, 8, …
Which is very similar to the sequence in part a. The pond would reach the same number of amoebas one day earlier, on February 9.

1-9. Trina's Sequence Machine

Trina has a sequence machine. She gives it a first number and the machine manufactures the next term, according to these rules:

Rule 1: If the number is less than 10, add 3 to get the next term.
Rule 3: If the number is equal to 10, subtract 5 to get the next term.
Rule 4: If the number is greater than 10, subtract 6 to get the next term.

a. If the first term is 9, what is the 100th term?

Solution: The terms form a pattern: 9, 12, 6, 9, 12, 6, 9, 12, 6, …
Every third term is 6. The 99th term would be 6, and the 100th term would be 9.

b. If the first term is 2, what is the 100th term?

Solution: The terms form a pattern: 2, 5, 8, 11, 5, 8, 11, 5, 8, 11, 5, 8, 11, …
The 3rd, 6th, … 99th terms are 8. So the 100th term is 11.

Follow up question: Work out the some sequences with different starting terms. Do you see some patterns?

1-10. Julio's Sequence Machine

Julio has a sequence machine. He gives it a first number and it manufactures the next terms according to these rules:

Rule 1: If the number is greater than 8, subtract 7 to get the next term.
Rule 2: If the number is less than or equal to 8, double it to get the next term.

Work out the sequences for various different starting numbers.

First Number	Julio's Sequence
0	0, 0, 0
1	1, 2, 4, 8, 16, 9, 2, 4, 8, 16, 9, 2, ...
2	2, 4, 8, 16, 2, ...
3	3, 6, 12, 5, 10, 3, ...
4	4, 8, 16, 9, 2, 4, ...
5	5, 10, 3, 6, 12, 5, ...
6	6, 12, 5, 10, 3, 6, ...
7	7, 14, 7, 14, 7, ...
8	8, 16, 9, 2, 4, 8, ...
9	9, 2, 4, 8, 16, 9, ...

What patterns do you see?
Each sequence is different, because it has a different first term, but are some of Julio's sequences similar to each other, or related to each other?
What's a good way to define *similar sequence* in the world of Julio's sequences?

Solution: The sequences that start with 1, 2, 4, 8, and 9 have similar behavior. The sequences that start with 3, 5, and 6 are similar. The sequence that starts with 0 is the only one that consists of only 0. The sequence that starts with 7 is similar to a sequence that starts with 14 or any multiple of 7.

One way to define similar is: two sequences are similar if they have the same "tail"—they might begin with different numbers, but after a finite number of terms, they settle down to the same pattern. According to this definition, the sequences that begin with 1, 2, 4, 8, and 9 are similar. The ones that begin with 3, 5, and 6 are similar, and the ones that begin with 7 and 14 are similar.

Follow up question: How many categories of Julio's sequence are there? Of the sequences that start with 0 to 9, there are four different categories: those whose tails are:

0, 0, 0, ...
2, 4, 8, 16, 9, ...
3, 6, 12, 5, 10, ...
7, 14, ...
Are there any more categories?

How To Deal With Sequence Problems

The following are only suggestions:

Learn the vocabulary:

- sequence
- consecutive
- term
- ellipsis

Look for patterns such as:

- adding
- subtracting
- multiplying
- perfect squares
- prime numbers
- odd/even numbers
- a steadily changing pattern
- use lists, charts, diagrams to help spot the pattern

There might be surprises such as:

- digits transposed
- sequence machines that have different patterns

2. Counting

Why Counting

Here are Professor Bear's reasons to for the second chapter to be about Counting:

- It's usually a new idea to people that Counting is a legitimate subject. It can be very entertaining to find out that Counting is more than it seems.
- Counting is a big part of any discrete mathematics (discrete as opposed to continuous—mathematics that involves separate, countable objects). Therefore, counting strategies are important in discrete Probability and Combinatorics.
- Probability is obviously far reaching and useful in modern statistical analysis.
- Combinatorics plays a large role in computer science, chip design, algorithm design.
- Counting is great for teaching problem solving because typically what's needed to solve a problem is a strategy rather than a single algorithm, and several different strategies can work.
- Often, within Counting problems are "how many ways…" questions that encourage divergent thinking not only to counting but to whatever is being counted.

Underpinnings of Counting

To understand this material you need some experience with arithmetic and be able to do one and two digit addition, subtraction, multiplication and division.

It's also helpful but not necessary to know fractions.

Outcomes

After doing this chapter, students will have more experience with strategies for counting, learn to recognize counting problems, and have a foundation they can apply to probability, statistics, and combinatorics.

Solutions: Notes and Vocabulary

Counting is probably the first thing you ever learned to do in math. It might seem like the easiest thing to do. But there are many different kinds of counting problems.

Here is one kind of problem: How many rectangles can you count in the following diagram?

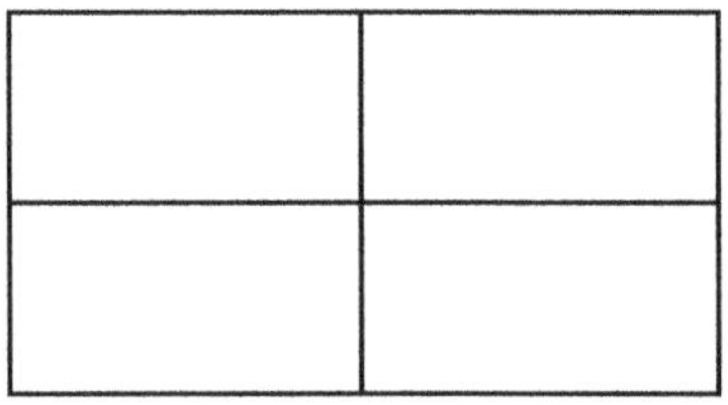

It looks like a diagram made up of four rectangles. But there are more, because for example, the top two rectangles together make up a fifth rectangle. How do you make sure you have counted all possible rectangles?

Think about it. A strategy is described on the next page.

Here is something else to think about. Here are two boxes of sixteen dots.

Box A

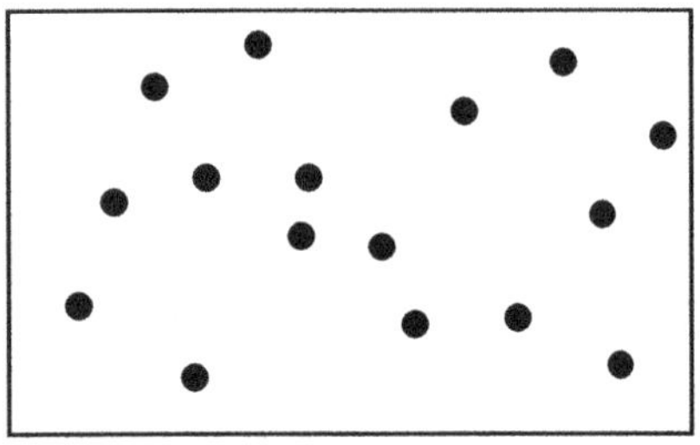

Box B

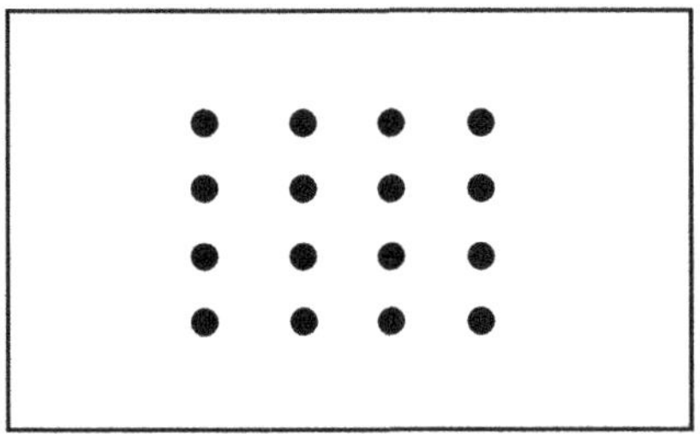

Which set of dots is easier to count, A or B? **B**

Write down some reasons for your answer.

Solution: The dots in box B are arrayed in order. It's easy to count them by fours. In the amount of time it takes to count the dots in Box A one-by-one, you could probably count the dots in Box B by fours twice, in two different directions. Box B gives you troubleshooting information—it tells you that your result should be even, and should be a multiple of four.

Here is a third kind of counting problem. Suppose you have a coin with a head on one side, a tail on the other. If you toss the coin twice, how many different combinations of heads and tails can you have?

Suppose the first toss gives you a head. The possibilities are: the second toss is also a head, and the second toss is a tail. We can list the two possibilities of head first like this:

HH and HT

Now, to count up all the other possibilities, they would each be the same except they would have a T first:

TH and TT

What this means is that you can count up the head-first possibilities first, and then double them, knowing that for every head-first possibility, there is a corresponding tail-first possibility. This method is an even better short cut if you try to count all the possible head-tail combinations for three coin tosses. Here are the head-first combinations:

HHH HHT HTH HTT

Without listing further, you can say that the total number of combinations is eight.

But just to make sure, here are all the other combinations:

THH THT TTH TTT

Make sure that there are no other combinations.

Here is a strategy for counting the rectangles in the diagram on the previous page.

Label the smallest rectangles. Every possible rectangle is made up of one or more smallest rectangles. Write down all the combinations.

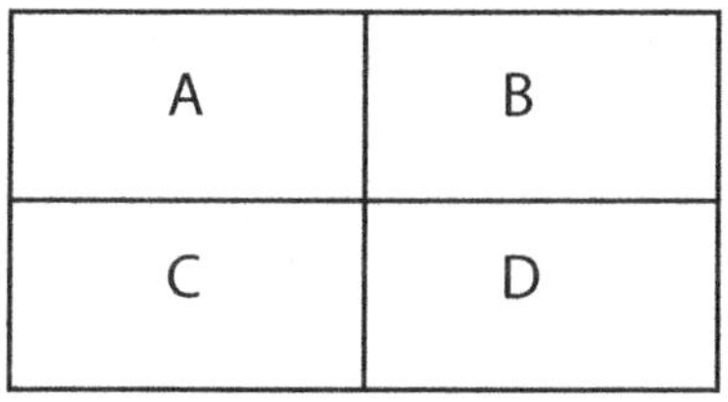

The possible rectangles are A, B, C, D, AB, AC, BD, CD, and ABCD for a total of nine.

Solutions: Problems

2-1. Counting Triangles

Different triangles can be traced using the lines in the figure given below. How many different triangles can be traced?

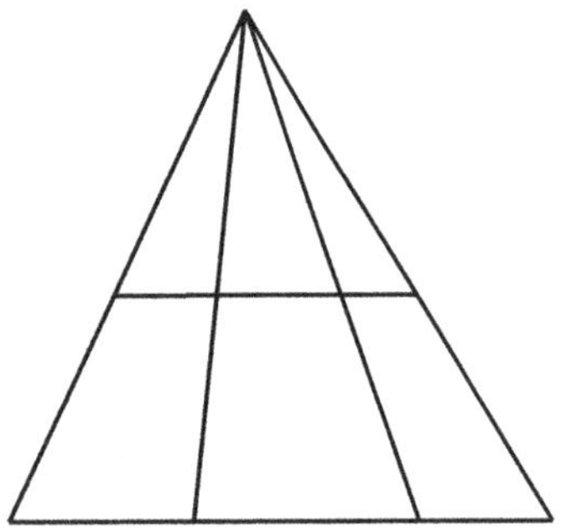

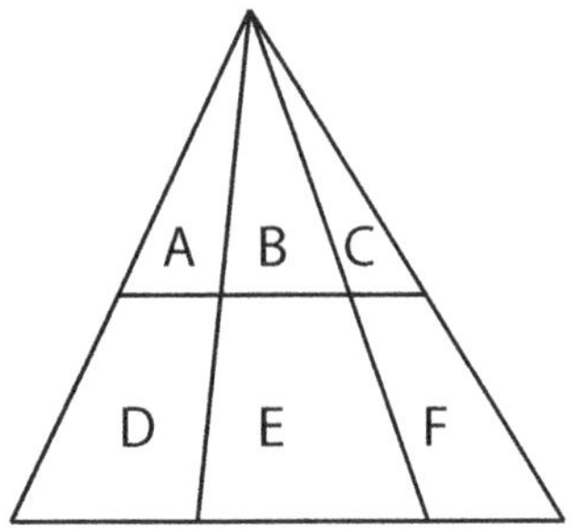

Solution: Using the suggested strategy of labeling the sub-areas, we get:
A, B, C, AD, BE, CF, AB, BC, ABDE, BCEF, ABC, ABCDEF
for a total of 12 triangles.

2-2. Five-Digit Palindromes

How many different five-digit palindromes are there?

Note: a palindrome is a number that reads the same forwards and backwards, such as 232 or 47788774 or 11611).

Solution: Solving this problem amounts to organizing the counting. There are ten possible digits, 0, 1, 2, 3, 4, 5, 6, 7, 8, and 9. Every five-digit palindrome is of the form ABCBA where:
A can be any digit from 1 to 9
B can be any digit from 0 through 9
C can be any digit from 0 through 9

I would start by looking at C. If C = 0, for example, the possible palindromes are all of the form AB0BA. There are nine possible values of A, and ten possible values of B, so there are ninety palindromes of the form AB0BA.

You might not believe this fact at first. It is very good to go through the exercise of convincing yourself, using one of these methods:

Let A=1. Then there are ten possible palindromes 1B0B1. Then let A=2, and so on, and it becomes clear that for each of the nine values of A, there are ten corresponding palindromes.

Look at a simpler problem. Suppose you want to count all two digit numbers of the form AB where A can be 1, 2, or 3, and B can be 0, 1, 2, 3, or 4. You'll see that the answer is 15, the product of 3 by 5.

Now it should be clear that if C=0, there are ninety possible palindromes. There are ten possibilities of C, so there are ten x ninety = nine hundred possible five-digit palindromes.

If anyone has trouble with this problem, you can first ask: how many 2-digit palindromes are there? How many 3-digit palindromes?

2-3. Ice Cream Choices

Pat is deciding on her ice cream sundae. She is allowed to have one flavor of ice cream, one syrup, and one topping. Her ice cream flavor choices are: chocolate, strawberry, vanilla. Her syrup choices are: hot fudge, caramel, raspberry, pineapple. Her topping choices are: nuts, sprinkles, gummy bears. How many different possible sundae combinations are there?

Solution: This problem is very like the previous one. The quick answer is 3 x 4 x 3 (3 kinds of ice cream, 4 types of syrup, 3 types of topping) which is 36. The challenge lies in recognizing that it is similar to the previous problem and the different possibilities of each component can be multiplied together.

2-4. Routes

Here is a map of Alice's house and Bernie's house. Alice lives on the corner of Snowy Street and Windy Street. Bernie lives on the corner of Foggy Street and Rainy Street. How many different routes are there from Alice's house to Bernie's house?

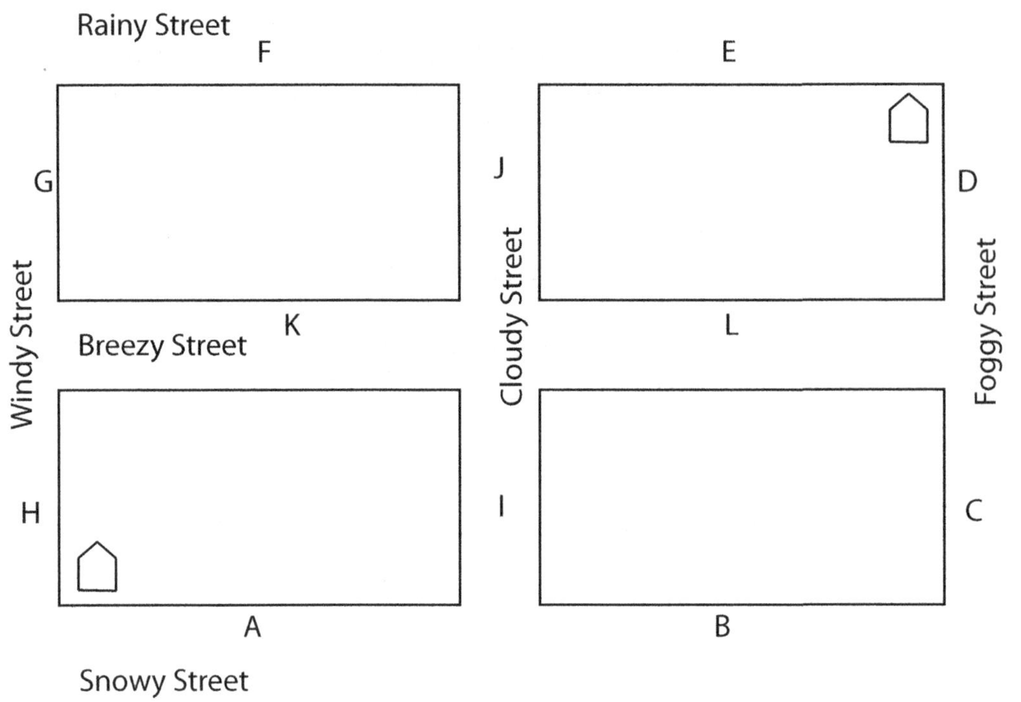

When you answer the first question, you might find that there are different types of routes. How many different *shortest* routes are there from Alice to Bernie's house?

Note to coach: The labeling of Alice's and Bernie's house is deliberately left out to encourage readers to translate the text to the picture.

Solution: This problem also lends itself well to the strategy of labeling. You can label the blocks and write down different paths.

The answer to the first question is that infinitely many paths exist, involving circling around and around Alice's block, for example.

There are 6 "shortest" paths, however, and that is probably the most meaningful answer to the question. The shortest paths are all 4 blocks long: ABCD, AILD, AIJE, HGFE, HKLD, HKJE.

2-5. Page Numbers

Agent Smiley numbers his pages of secret notes consecutively, beginning with 1. This requires a total of 234 digits. How many pages of secret notes does Agent Smiley have?

Solution: This problem involves organizing the counting of page numbers. There are:
nine one-digit page numbers, using up 9 digits

From 10 to 19 are 10 numbers, using up 20 digits; from 10 to 99 there are nine groups of 20 digits, using up 180 digits
Up to page 99, 189 digits are used
Starting at 100, each number uses 3 digits. To find out how many 3-digit numbers there are, subtract 189 from 234 to get: 45 digits used for 3-digit numbers.
45 divided by 3 is 15. The 3-digit numbers are 100 through 114.
Agent Smiley has 114 pages of notes.

Note to coach: this solution assumes that one "page" is one side of a physical page. If the student assumes that "page" means physical piece of paper, the solution would be 57.

2-6. Cover Pattern Counting

The figure on the cover of this book shows 4 patterns of squares.

a. How many sides of squares are on the first (upper left) pattern?

Solution: 5x4 = 20

b. How many sides of squares are on the second (upper right) pattern?

Solution: 5x5x4 = 100

c. How many sides of squares are on the third (lower left) pattern?

Solution: 5x5x5x4 = 500

d. How many sides of little tiny squares are on the fourth (lower right) pattern?

Solution: 5x5x5x5x4 = 2500

2-7. Distinct Numbers

How many distinct 4-digit counting numbers can you construct from the digits 3, 0, 0, 7?

Solution: This problem has some logical follow-through, since the zeros can't be used at the beginning of the number. The possibilities are:

- 3XXX or 7XXX
- Once you count the possibilities beginning with 3, you can double that number to include the possibilities that begin with 7
- The possibilities that begin with 3 are: 3700, 3070, 3007

The solution is therefore 6.

2-8. Permutations

Count Rivula has three precious beads: a topaz, an emerald, and a sapphire. The beads can be strung in different order on a necklace that fits around the Count's ruffled collar.

a. How many different bead orders are possible?

Solution: It's worthwhile to label the beads as T for topaz, E for emerald, and S for sapphire, and the systematically write down the different permutations:
TES, TSE, ETS, EST, STE, SET
The solution is 6. Looking at the permutations above, you can see there are some shortcuts to the problem. There are 2 permutations that start with T. There are 3 different letters to start with, and 2 permutations for each starting-letter, therefore 3x2 = 6 permutations.

Note to Coach: Notice the difference between this problem and the palindrome problem. In the palindrome problem, digits can be repeated. In this problem, we have a fixed set of beads and all we are counting are the *rearrangements*. Permutations are rearrangements or changed orders of a fixed set of objects.

b. Countess Rivula gives the Count an additional bead, a ruby. How many different bead orders are possible using all 4 beads?

Solution: Now we have T, E, S, and R. Using the shortcut thinking of part a, we know that if T is the starting bead, there are 6 permutations of the remaining beads. There are 4 possible starting beads, so the solution is 6x4 = 24.

This kind of thinking breaks down nicely as follows: If you have two beads, you have 2 permutations; if you have 3 beads, you have 3x2 permutations. If you have 4 beads, you have 4x3x2 permutations. And so on.

Note to coach: There is nothing wrong with letting students write down all the permutations without taking shortcuts. It might help them appreciate the shortcut, or work out the shortcut. They might be suspicious of any shortcut without going through the process of writing all the permutations down. In that case, it's important to encourage organized ways of listing the permutations, which will help patterns to emerge.

2-9. Assortments

Henry is allotted 6 cookies from a jar. The jar contains the following types of cookie: chocolate chip, peanut butter, and oatmeal. There are at least 6 cookies of each type in the jar. How many different assortments of cookies could Henry choose?
Note: The following two assortments of cookies are different:

- 2 chocolate chip, 2 oatmeal, 2 peanut butter
- 2 chocolate chip, 1 oatmeal, 3 peanut butter

These two assortments of cookies are the same:

- 3 oatmeal, 2 chocolate chip, 1 peanut butter
- 2 oatmeal, 2 chocolate chip, 1 peanut butter, 1 oatmeal

Solution: Depending on how the students think about the problem, there can be confusion about what constitutes the "same" assortment. Some students will look only at the end result of cookie counts, and not the different ways to get there.

Let chocolate chip = C, peanut butter = P, and oatmeal =0. A good way to organize the counting is to follow what happens with C, adding more and more O and P:

CCCCCC
CCCCCO, CCCCCP
CCCCOO, CCCCOP, CCCCPP
CCCOOO, CCCOOP, CCCOPP, CCCPPP
...
OOOOOO, OOOOOP, OOOOPP, OOOPPP, OOPPPP, OPPPPP, PPPPPP

If you continue, you see that a pyramid develops, with the last line being the one with zero C, and just O and P. The sum of the assortments is the 7th triangular number: 1 + 2 + 3 + 4 + 5 + 6 + 7 = 28.

2-10. Robot Moves

Arlene makes a little robot that moves on a chessboard. The robot can move forward, backward, left, and right, but not diagonally. 5 moves can be programmed at a time. Depending on the sequence of moves, the robot ends up on different destination squares. How many different squares can the robot reach after 5 moves, starting from the upper left?

Solution: A pattern to notice here is that starting from the white square on the upper left, every odd-numbered step lands on a black square, and every even-numbered step lands on a white square. The possible landing squares are shown above.

There are many ways to demonstrate the solution above. One way is to show how each square can be reached in 5 moves: the brute force approach. Some students enjoy brute force. Another way is to draw a diagonal from top left to bottom right, and show how the squares can be reached on one side of the diagonal. By symmetry, the squares on the other side of the diagonal can be reached.

How To Deal With Counting Problems

The following are only suggestions:

Learn the vocabulary:

- Digit
- Permutation
- Rearrangement
- Combination
- Assortment
- Palindrome
- Brute force

Look for patterns such as:

- Symmetry: divide and conquer
- Multiplication

There might be surprises such as:

- There might be more counts than you think

3. Factors And Primes

Why Factors And Primes
Here are Professor Bear's reasons to for the third chapter to be about Factors And Primes:

- Factors are a big subject in elementary school math.
- Prime numbers are important in cryptography and number theory.
- The most important theorem in arithmetic, the Fundamental Theorem of Arithmetic, relates to prime numbers.

Underpinnings of Factors And Primes
To understand this material you need experience with multiplication and division.
It's good to know the multiplication facts well.
It's also helpful but not necessary to know fractions.

Outcomes
After doing this chapter you'll have more experience with several divisibility rules and prime numbers.

Solutions: Notes and Vocabulary

The **factors** of a whole number are the numbers that divide it evenly. In other words, if you divide a number by one of its factors, the remainder is zero.

The factors of 6 are 1, 2, 3, and 6.

What are the three factors of 9?   1, 3, 9

Some numbers, such as 12, have many factors. Name all the factors of 12:

1, 2, 3, 4, 6, 12

Name all the factors of 13:   1, 13

13 is an example of a **prime number**. The only factors of prime numbers are themselves and 1.

True or False:

	Statement	T or F?	Example or explanation
1	If 3 is a factor of a number, you can count up to that number by threes.	**T**	**"3 is a factor of a number" is the same as "the number is a multiple of 3" or "the number can be grouped in 3s".**
2	3 is a prime number.	**T**	**The only factors of 3 are 1 and 3.**
3	Find 3 consecutive numbers with a remainder of 2 when you divide by 5. These numbers are all even.	**F**	**Example: 12, 17, 22. 17 is odd.**
4	If 2 is a factor of a number, then 4 is also a factor of that number.	**F**	**Example: 2.** **Another example: 10**
5	If 6 is a factor of a number, then 3 is also a factor of that number.	**T**	**If 6 is a factor of A, then A = 6xB.** **6xB = 3x2xB = 3x(2xB) =** **A = 3xC, where C = 2xB.** **So 3 is a factor of A.**
6	Some numbers have factors that are prime numbers.	**T**	**True for all the prime numbers. 6 has two prime factors, 2 and 3.**
7	Some numbers have factors that are not prime numbers.	**T**	**One factor of 20 is 4, which is not prime.**
8	Every number has at least one factor that is a prime number.	**T**	**(Keep in mind the number here is whole). If the number is prime, then it has one factor which is a prime number. If the number is not prime, then it has non-prime factors. Each non-prime factor is either prime, or has smaller factors that are not prime. Continue this process, and eventually a factor has to be prime.**

Note to Coach: The last statement, 8, can be explained more fully. It's worthwhile to explore what factors look like for any given number. Factors are distributed around the square root of the number. If the number of factors is even, half are bigger than the square root, and half are smaller than the square root. If the number of factors is odd, the square root is the middle factor and of the remaining factors, half are less than the square root and half are greater.

What this means is that the process described in statement 8 has to come to an end. If you look at all of the factors of a number N, some will be less than the square root of N, which is less than N. Let F1 be one of these factors of N, less than the square root of N. If F1 is not prime, then F1 has factors, some of which are less than the square root of F1. If none of F1's factors are prime, keep going. It would be impossible for this process to go on forever, because each time the factors get smaller:

N > square root of N > F1 > square root of F1 > F2 > square root of F2 ...

If the process went on forever without turning up a prime factor, N would have to be infinite, which it isn't.

This argument is quite subtle and youngsters probably won't be able to write it down, but they can think about it.

Two obvious prime numbers are 3 and 7.

Name all the prime numbers between 10 and 20: **11, 13, 17, 19**

What is the smallest prime number? **2**

You might think that the smallest prime number is 1. Its only factors are itself and 1. But the number 1 is not accepted as a prime number. The main reason for this has something to do with the Fundamental Theorem of Arithmetic.

A **theorem** is a mathematical truth that has been proved by at least one mathematician. Many people have proved the Fundamental Theorem of Arithmetic.

Fundamental Theorem of Arithmetic

Every whole number greater than one is either prime or a product of prime numbers. The prime factors of a non-prime number are unique, meaning that if you arrange the prime factors in order of size, there is only one way to do it.

Examples:
7 is prime.
6 is a product of primes: 2 x 3
4 is a product of primes: 2 x 2
11 is prime.

12 is a product of primes: 2 x 2 x 3
12 is a product of many factors, such as 2 x 6 or 3 x 4, but there is *only one way* to express 12 as a product of prime numbers arranged from lowest to largest.

2 x 2 x 3 is called the **prime factorization** of 12.

Prime factorizations are useful for solving problems, because the prime factors of a number are the building blocks of all of the factors of the number. The number 12, for example, has the non-prime factor of 6. But 6 is the product of two prime factors, 2 and 3. So the 2 and the 3 within 12 build up the factor of 6.

Here is a problem that uses the prime factorization idea:

The product of two consecutive whole numbers is 210. What are these two numbers?

Solution: the prime factorization of 210 can be rearranged into any of the factors of 210. So the prime factorization gives us a place to start. 210 is divisible by 2, 3, 5, and 7 (check the divisibility rules).

The prime factorization of 210 is turns out to be 2 x 3 x 5 x 7. We can see that these numbers can be rearranged into 2 x 7 x 3 x 5, which is 14 x 15. Therefore the solution of our problem is: the two consecutive numbers are 14 and 15.

Perhaps you can now see why it's not a good idea to make 1 a prime number. What would happen to the prime factorization of a number if 1 were allowed to be a prime number?

Solution: If 1 were allowed to be a prime number, then many different "prime" factorizations of a number would be possible:
6 = 1 x 2 x 3 = 1 x 1 x 2 x 3 = 1 x 1 x 1 x 2 x 3 , etc.

Divisibility Rules

A divisibility rule tells you whether or not a number is divisible by a certain number. Here are some divisibility rules.

Factor	Divisibility Rule
2	Number ends in 0, 2, 4, 6, or 8.
3	The sum of the digits is divisible by 3. Name a 5-digit number divisible by 3:
4	The last 2 digits have to be divisible by 4.
5	Number ends in 0 or 5.
7	1. Double the last digit. 2. Subtract the double from the remaining digits and see if the

	resulting number is divisible by 7. Example: is 406 divisible by 7? 6 x 2 = 12, and 40 - 12 = 28 which is divisible by 7. So 406 is divisible by 7.
9	The sum of the digits is divisible by 9. Name a 4-digit number divisible by 9:
10	Number ends in 0.
100	Number ends in 00. Name a 6-digit number divisible by 100:

Numbers such as 4 and 10 are **composite** numbers (whole numbers that are not primes). To be divisible by 10, a number has to be divisible by both 2 and 5. Write down some divisibility rules for these composite numbers:

Factor	Divisibility Rule
6	Divisible by both 2 and 3.
15	Divisible by both 3 and 5.

3-1. Commutativity

What is the value, in simplest terms:

a. (20 x 24 x 28 x 32) ÷ (10 x 12 x 14 x 16) = **16**

Note to Coach: This is a good opportunity to write down simpler problems and have them solve them. You also have the option to teach the students the vocabulary word "commutative" if it would be useful to them.

There are many ways of course to solve this problem. It's good for students to show different rearrangements of the numbers that lead to a solution.

b. (26 x 12 x 28 x 15) ÷ (13 x 24 x 14 x 30) = **1**

3-2. Factor Counts

The number 6 has four factors: 1, 2, 3, and 6.

a. Which number has more factors, 24 or 28?

Solution: 24 has the factors: 1, 2, 3, 4, 6, 8, 12, 24
28 has the factors: 1, 2, 4, 7, 14, 28, so 24 has more.

b. Which number has more factors, 31 or 33?

Solution: 31 is prime. 33 is not prime, since 3 is a factor, so it has more factors. Explicitly, the factors of 31 are 1 and 31. The factors of 33 are 1, 3, 11, and 33.

3-3. Remainders

What positive whole numbers have a remainder of 4 when you divide them into 32?

Solution: The key number here is 28, which is 4 less than 32. All possible solutions would be factors of 28: 1, 2, 4, 7, 14, 28. 1, 2, and 4 are not solutions because they divide evenly into 32. The solutions are 7, 14, and 28.

3-4. Large Power Of 2

2^{2012} means the number 2 multiplied by itself 2012 times. It is a very large number. What is its ones digit?

Solution: The ones digit of powers of 2 follow a pattern: 2, 4, 8, 6, 2, 4, 8, 6, ... and so on. Every 4th power of 2 has a ones digit of 6. 2012 is a multiple of 4, so its ones digit is 6.

3-5. Divisibility

How many positive whole numbers less than 100 are divisible by 6 or 21 or by both?

Solution: This question has two "or" words in it, so there are a lot of solutions: all the multiples of 6 less than 100 (16), all the multiples of 21 (4, but two of them are multiples of 6 already). This brings the total to 18.

3-6. Ending Zeros

The number 25! (25 factorial) is equal to 1 x 2 x 3 x ... x 23 x 24 x 25 and is a very large number. It ends with a certain number of 0 (zero) digits.

How many zeroes are at the right-most end? Give reasons for your answer (explain why there are no fewer zeroes, and no more zeroes, than your answer).

Solution: The only way you get a zero at the end is from a factor of ten, or 2x5. Looking at:

1 x 2 x 3 x 4 x 5 ... x 10 x ... x 15 x ... x 20 x 21 x ... x 25

We can underline the numbers where there are factors of five:

$1 \times 2 \times 3 \times 4 \times \underline{5} \ldots \times \underline{10} \times \ldots \times \underline{15} \times \ldots \times \underline{20} \times 21 \times \ldots \times \underline{25}$

The number of five-factors is: 6 (there are two fives in 25).

There are at least 6 twos in the product, because (2 x 4 x 8 = 2 x 2 x 2 x 2 x 2 x 2).
So the number of zeros is 6.

3-7. Consecutive Factors

a. The product of two consecutive numbers is 1806. What are these two consecutive numbers?

Solution: The decomposition of 1806 = 2 x 903 = 2 x 3 x 301 = 2 x 3 x 7 x 43

2 x 3 x 7 = 42, so 1806 = 42 x 43

Alternatively: the last digit of 1806 is 6. Two consecutive factors that multiply to 6 are 3 and 2. The question becomes, what two-digit numbers ending in 2 and 3 multiply to 1806? 22 and 23 are too small, 32 and 33 too small, 42 and 43 work. (Notice that 1806 is between 40x40=1600 and 50x50=2500).

b. The product of two consecutive numbers is 1260. What are these two consecutive numbers?

Solution: 1260 = 2 x 630 = 2 x 7 x 90 = 2 x 2 x 45 x 7 = 2 x 2 x 3 x 3 x 5 x 7
There are many pairs of factors of 1260 but the consecutive pair are: 35 x 36

Alternatively: the last digit of 1260 is 0. What two consecutive numbers multiply together to form a number whose last digit is 0? The only combinations are 4 and 5 and 5 and 6. 1260 is between 900 (30 x 30) and 1600 (40 x 40) so the numbers are either 34 and 35 or 35 and 36. A simple computation leads to 35 and 36.

3-8. Factors Of Consecutive Numbers

Name 3 consecutive numbers less than 100, where the smallest is divisible by 7, the next number is divisible by 5, and the largest number is divisible by 3.

Solution: The coarsest "sieve" here is divisibility by 7. So one way to begin is by looking at multiples of 7, and consecutive numbers. The first multiple of 7 with a following number divisible by 5 is 14. But the next number after 15 is 16, which is not divisible by 3. The next candidate is 49, followed by 50, which is divisible by 5. The next number, 51, is divisible by 3.

3-9. Remainders With Apples

Jill has some apples. If she puts them in groups of 5, she has 3 left over. If she puts them in groups of 7, she has 2 left over. What is the least number of apples she could have?

Solution: Multiples of 7 with remainder 2: 9, 16, 23, 30, 37, …
Multiples of 5 with remainder 3: 8, 13, 18, 23 –Aha! The solution is 23.

3-10. 12345

How many different five-digit numbers can you make using the digits 12345?

Solution: This problem is a throwback to the Counting chapter. The question is the number of permutations of 5 digits. The number of permutations is 5x4x3x2x1 = 120.

If the first digit is 1, there are 24 permutations of the remaining 4 digits. Therefore there are 5x24 permutations altogether.

How many of these numbers are prime?

None of the permutations are prime, because the sum of the digits is 15, which is divisible by 3, hence all of the numbers are divisible by 3.

How To Deal With Factors and Primes Problems

The following are only suggestions:

Learn the vocabulary:

- Factor
- Prime
- Prime factorization
- Remainder
- Powers of factors
- Divisibility
- Perfect squares, perfect cubes

Look for patterns such as:

- Last digits: when factoring a large number, what are the factors of the ones digit
- Divisibility rules

There might be surprises such as:

- A problem might seem hard when you read it, but remember they are all related to multiplication, division, remainders, and prime numbers.
- A large number can be factored quite quickly if you are able to divide out small prime factors.

4. Related Numbers

Why Related Numbers

Here are Professor Bear's reasons to for the fourth chapter to be about Related Numbers:

- Some people see the material in this chapter as "algebra word problems". However, many of these problems can be solved using pre-algebra techniques.
- It's valuable to have experience solving these problems without the symbolic baggage of algebra.
- It's also valuable to motivate learning algebra by doing these kinds of problems.
- These problems stand at the intersection of math and reading.
- This chapter is a little more challenging and uses some of the ideas in later chapters, Logic in particular. But the chapter is located early in the book for the following reason: if students are using this book to practice for some kind of math contest, it's good to have more time to practice and think about these kinds of problems.

Note to Coach: Avoid teaching algebra to elementary school students in order to solve Related Numbers problems. Do not rob them of the chance to experiment with pre-algebraic techniques and logic to solve these problems. It's very important to have a foundation and motivation for learning algebra, in order to appreciate the powerful techniques they will learn in the future.

Underpinnings of Related Numbers

To understand this material you need some experience with arithmetic and be able to do one and two digit addition, subtraction, multiplication and division.

It's also helpful but not necessary to know fractions. It helps to read well and carefully.

Outcomes

After doing this chapter students will be able to recognize several different types of word problems, and apply a few different techniques to solve them.

Solutions: Notes and Vocabulary

There are many problems about related amounts of things. Some examples:

- Sarah has 2 marbles. George has 5 marbles more than Sarah. How many marbles does George have?
- Marissa drove for 2 hours at 40 miles per hour. How far did she travel?
- Chu can paint 1 garage door in 1 hour. Sven can paint 1 garage door in 3 hours. How long would it take for them to paint 4 garage doors if they work together?
- In the morning Jose spent half his money on a package of batteries, and then at noon he spent two-thirds of what was left on a sandwich. Now he has 2 dollars. How much money did he have to start with?

You might be familiar with the term "word problem." These problems are all word problems. All of these word problems have something in common: they relate one number to another number.

- Sarah's number of marbles is related to George's number of marbles.
- Marissa's driving time and her speed are related to how far she traveled.
- Chu's painting time and Sven's painting time are related to how long it would take them to paint something together.
- Jose's number of dollars, after buying a sandwich and some batteries, is related to how much money he had in the morning.

The problems above can be solved in many different ways.

Technique: Working Backwards

The problem of Jose's money is ideal for solving by working backwards.
At the end of the problem, he has 2 dollars. He had just spent two-thirds of his money on a sandwich and 2 dollars was left, so 2 dollars is the remaining one-third.

Just before he bought the sandwich, he must have had 6 dollars. The 6 dollars is the half left after he bought batteries, so before he bought the batteries, he must have had 12 dollars. And that is how much money he had to start with.

Technique: Read With Care and Use Logic

This basic technique is vital for all related number problems. It's all you need to solve the first problem: *Sarah has 2 marbles. George has 5 marbles more than Sarah. How many marbles does George have?*

In a problem like this, you read for the information that will help you solve the problem. You could ask yourself: What do I know?

How many marbles does Sarah have? **2**.
Who has more marbles, George or Sarah? George—so he must have more than 2 marbles.

How many more marbles does George have than Sarah? **5**.
The words "more than" mean that to get George's number of marbles, you have to add 5 to Sarah's number. That means the answer is 7.

Technique: Use the Speed, Time, and Distance Relationship
There is a simple relationship between speed, time, and distance traveled. This is what you use to solve the next problem: *Marissa drove for 2 hours at 40 miles per hour. How far did she travel?*

The meaning of "forty miles per hour" is: in 1 hour, 40 miles are traveled. In 2 hours, 40 + 40 miles are traveled. If Marissa drove for 2 hours at 40 miles per hour, she traveled 80 miles.

You might find it useful to remember a formula: the product of the time and the speed is the distance traveled.

Here is another problem that uses the speed, time, and distance relationship: *Fred ran from home to the town of Hayfield, a distance of 12 miles, in 3 hours. What was Fred's average speed?*

The average speed is the total distance divided by the total time. In the case of this problem, Fred's average speed is:

12 miles ÷ 3 hours = 4 miles per hour.

The average speed is the speed that Fred would travel if he traveled at the exact same speed during all 3 hours.

The last simple problem of this type asks for the time: *How long does it take to fly a distance of 1000 miles in an airplane that travels 500 miles per hour?*

The travel time is the total distance divided by the speed. The answer here is 2 hours.

Technique: Use the Rate, Time and Job Relationship
This relationship is the same as the speed, time and distance relationship; it just involves different words. Instead of speed, these problems have a rate; instead of distance traveled, a job is done.

Instead of:
Distance = speed multiplied by time and time = distance divided by speed

We have:
Number of jobs = rate multiplied by time, and time = number of jobs divided by rate

We use the rate, time and job relationship to solve this problem: *Chu can paint 1 garage door in 1 hour. Sven can paint 1 garage door in 3 hours. How long would it take for them to paint 4 garage doors if they work together?*

This problem presents us with two rates. Chu's rate of door painting is 1 door per hour. Sven's rate of door painting is 1 door per 3 hours, or 1/3 of a door per hour.

The problem asks how long it would take Chu and Sven to paint 4 garage doors if they work together. There are several ways to look at this problem.

One way is to ask: what happens after 1 hour, if Chu and Sven work together? What happens in 2 hours? 3 hours? Is there a pattern? Try this out. Write down what happens:

After 1 hour: 1 door plus 1/3 of a door are painted.

After 2 hours: 2 doors plus 2/3 of a door are painted.

After 3 hours: 3 doors plus 3/3 of a door are painted—4 doors.

What do you think the solution is now? **3 hours**

Another approach: If Chu's rate is 1 door per hour, and Sven's rate is 1/3 door per hour, what is their combined rate (the rate for the two of them working together)? Think about it. You'd expect their combined rate to be bigger than each of their single rates.

The number of jobs to be done in this problem is 4 (4 doors to paint). Using our rate/time/jobs formula:

Time = number of jobs divided by rate = 4 divided by (combined rate)
See if this method gives you the same answer as the first method.

The combined rate is: 1 + 1/3 door per hour, which is 4/3 door per hour.
4 divided by 4/3 = 4 x ¾ = 3 hours.

Technique: Balancing

The idea of a balancing scale can help solve a lot of related number problems. Here's an example: *Mary the baker wants to figure out how much it costs to make a loaf of bread, and how much it costs to make a roll. She knows that 30 cents worth of supplies makes 1 loaf of bread and 5 rolls. And she knows that 60 cents worth of supplies makes 4 loaves of bread.*

We can pretend we have a balance that balances money. If 4 loaves of bread costs 60 cents, we can show this on a balance:

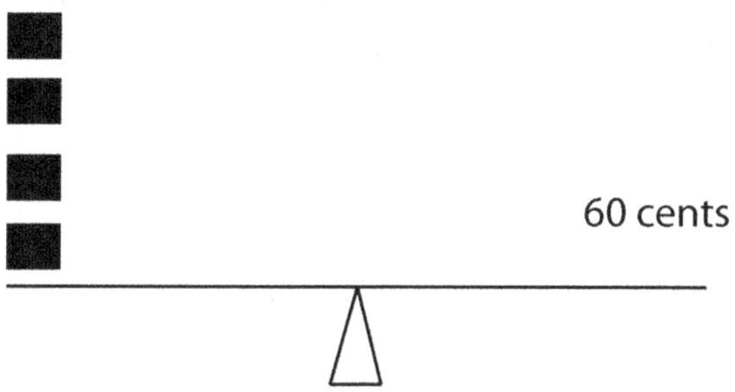

If we take away 2 loaves of bread from the left side, what do we take away from the right side in order to balance the scale? **30 cents**

Use this idea to figure out the cost of 1 loaf of bread: **15 cents**

Now, 1 loaf of bread and 5 rolls balances with 30 cents. Is there a way to balance *loaves and rolls* with 60 cents? Fill in the sketch:

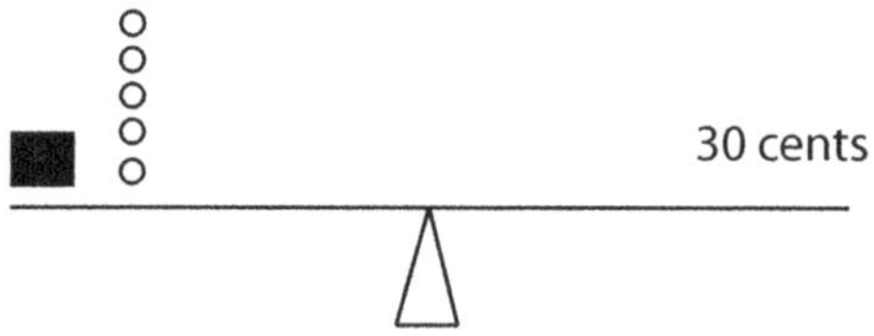

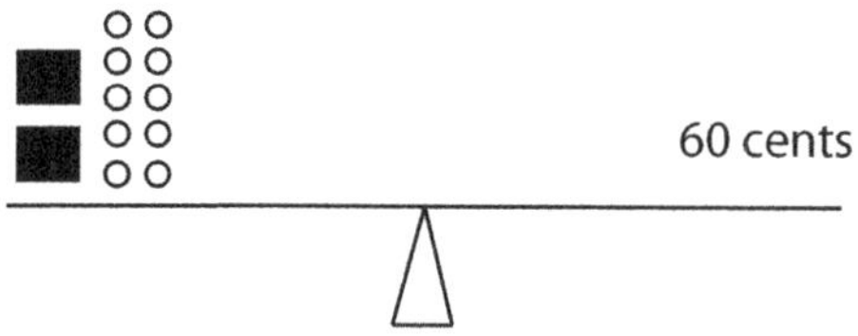

Now you can keep balancing to figure out the cost of 1 roll.

If 1 loaf and 5 rolls balance 30 cents, then 2 loaves and 10 rolls balance 60 cents. We also know that 4 loaves balance 60 cents. Therefore, 4 loaves balances 2 loaves and 10 rolls.

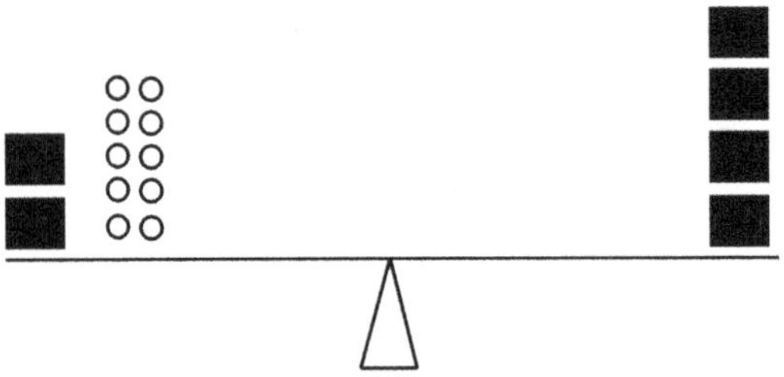

If you take the same things away from each side of the balance, the balance is still kept. Sketch what happens to the above balance if you take 2 loaves away from each side.

Solution:

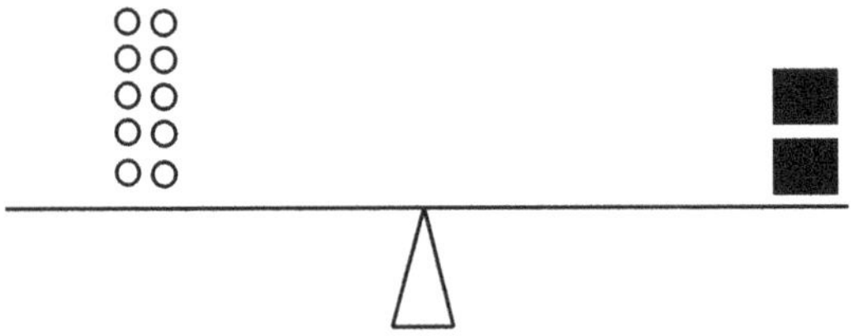

Next, sketch what happens to the balance if you take half of each side away.
Solution:

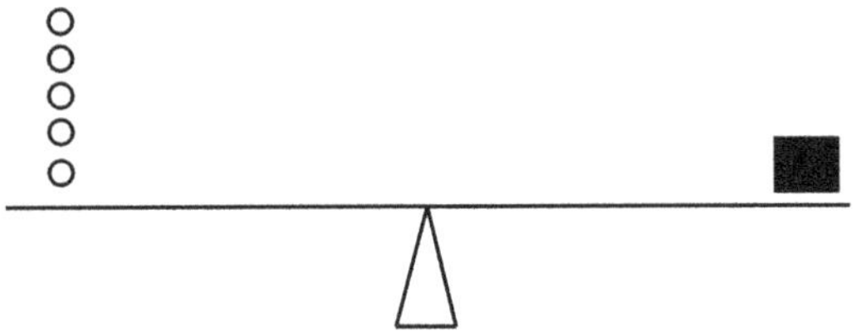

Re-sketch the previous balance, replacing the loaf with the amount of money it is worth. Solution:

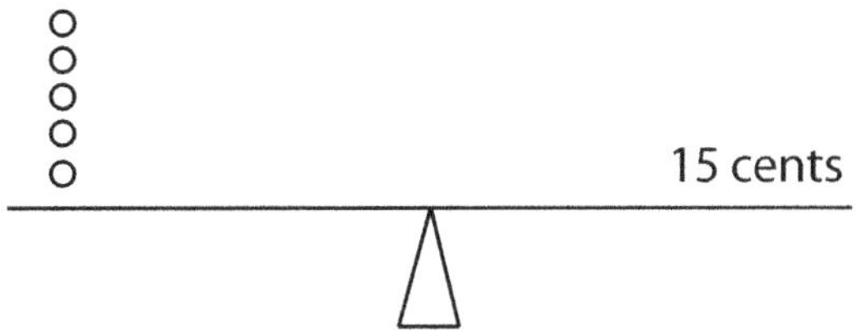

How much money does 1 roll cost? **3 cents**

Technique: Variables and Equations
A variable is a symbol for a quantity. A variable gives you a short way to write down a quantity. For example:

Herman the centipede has twice as many legs as Sandra the caterpillar.

If we let the variable X stand for the number of legs Sandra has, then the number of legs that Herman has is X + X.

In math, we say: Let X = the number of legs that Sandra has.
We can also say: Let Y = the number of legs that Herman has.

Then we can write an equation: X + X = Y

X + X means "two X's" and in math, we write that as 2X.

How do you think we write X + X + X? 3X

The equation relating Herman's and Sandra's legs can be written: 2X = Y

If Sandra has 10 legs, how many does Herman have? **20**

If Herman has 14 legs, how many does Sandra have? **7**

If you have a problem like the Loaf/Roll problem, you can use variables and equations. In fact, the equation technique and the balancing technique are the same. Instead of a balance, you write an equation. Every diagram of the balance can be converted to an equation. For example:

Balance	Equation
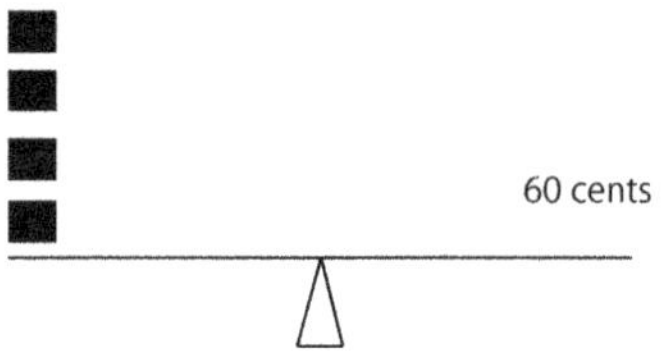	Let L = cost of 1 loaf L + L + L + L = 60 or: 4L = 60

Equations work like balances. Just like on a balance, you can take away half from both sides of an equation and you still get an equation.

If: L + L + L + L = 60 = 30 + 30
Then, taking half away from each side: L + L = 30

Another example from the previous problem:

Balance	Equation
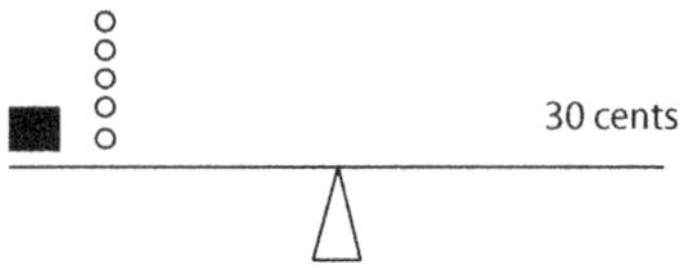	Let L = cost of 1 loaf Let R = cost of 1 roll L + R + R + R + R + R = 30 or: L + 5 R = 30

You can double equations:
If: L + 5R = 30
Then, doubling: L + 5R + L + 5R = 30 + 30
When you add, you can rearrange the items being added:
L + L + 5R + 5R = 30 + 30
And you can add up matching items:
2L + 10R = 60

Try converting this balance to an equation:

Balance	Equation
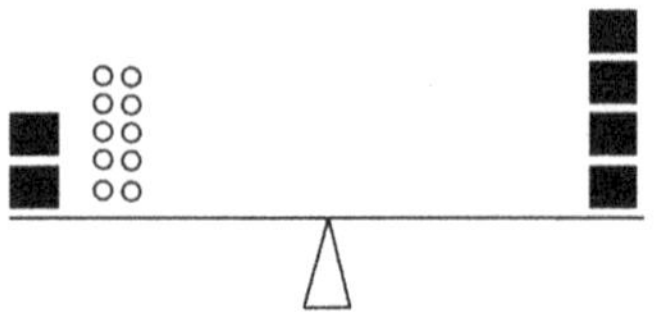	Let L = cost of 1 loaf Let R = cost of 1 roll 2L + 10R = 4 L

Try solving the Loaf/Roll problem entirely with equations, following the solution of the balances.

Solution: L + 5R = 30, therefore 2L + 10R = 60

2L + 10R = 4L, so 4L = 60

Grouping into 4 groups, we get
L + L + L + L = 15 + 15 + 15 + 15 and L = 15.

If L=15, then L + 5R = 15 + 5R = 30

Taking 15 from both sides, we get 5R = 15

Grouping into 5 groups, we get
R+R+R+R+R = 3 + 3 + 3 + 3 +3 and R = 3.

4-1. Balancing Fruit

One apple weighs as much as a banana and a tangerine together.

a. Draw a balance scale in a balanced position with an apple on one side.

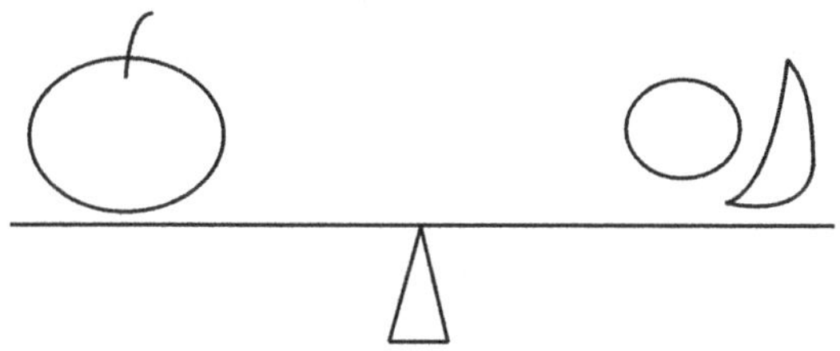

Answer the following questions, if possible. Some of these questions cannot be answered—there isn't enough information about all the fruits. If a question cannot be answered, say so and think about why.

b. Does the apple weigh more than the banana?

Solution: Yes, if the tangerine has weight.

c. Does the banana weigh more than the tangerine?

Solution: Not enough information to tell.

d. Does the apple weigh more than two bananas?

Solution: Not enough information.

e. Does the tangerine weigh less than the apple?

Solution: Yes, if the banana has weight.

f. Does the banana weigh less than the apple?

Solution: Yes, if the tangerine has weight.

g. Does the apple weigh more than two tangerines?

Solution: Not enough information.

h. Do two tangerines weigh less than two apples?

Solution: Yes, if the banana has weight.

4-2. Marbles

Shirley gave half of her marbles to Harry. Harry gave 3 marbles to Dion, and one-third of what he had left to Sharon. If Sharon has 4 marbles, how many did Shirley have?

Solution: Classic working backwards problem: Sharon has 4 marbles, which is one-third of what Harry had. Therefore Harry had 12. Harry gave 3 marbles to Dion, so Harry started with 15. This is half of what Shirley had, so Shirley had 30 marbles.

4-3. Car Factories

There are 3 factories in the province of Atheneum. The factory in Smallville builds 2 cars per month. The factory in Bigville builds 4 cars per month, and the factory in Rightville also builds 4 cars per month.

a. How long does it take the Smallville factory to build 10 cars?

Solution: Job = 10 cars; Rate = 2 cars/month
Time = Job divided by Rate = 5 months

b. How long does it take the Bigville factory to build 10 cars?

Solution: Job =10 cars; Rate = 4 cars/month
Time = Job divided by Rate = 2 ½ months

c. How many cars per month are built in the entire province of Atheneum?

Solution: In one month, Smallville builds 2 cars, Bigville builds 4 cars, and Rightville builds 4 cars, for a total of 10. The rate for Atheneum is 10 cars/month.

d. How long would it take for Atheneum to produce 200 cars?

Solution: Job =200 cars; Rate = 10 cars/month
Time = Job divided by Rate = 20 months

4-4. Chocolates And Lollipops

4 chocolates and 1 lollipop cost 48 cents.
1 chocolate and 4 lollipops cost 72 cents.
What is the cost of 1 chocolate and 1 lollipop?

Solution: This is a good problem for drawing a diagram or using some representative manipulative for chocolates and lollipops and moving them around. At some point, a key insight is that:

4 chocolates and 1 chocolate and 1 lollipop and 4 lollipops =
5 chocolates and 5 lollipops, costing
48 + 72 = 120 cents.

Dividing 120 into 5 groups gives the answer: 1 chocolate and 1 lollipop cost 24 cents.

4-5. Forks, Spoons, And Steel

From 4 ounces of steel, an artisan can make 1 fork and 1 spoon.
From 10 ounces of steel, the artisan can make 3 forks and 1 spoon.

a. How many ounces of steel go into 1 fork?

Solution: Like the previous problem, this is a good problem to draw diagrams of the forks and spooks or use real ones and move them around. At some point a key insight would be:
3 forks and 1 spoon take away 1 fork and 1 spoon is equivalent to:
10 ounces take away 4 ounces.

This means that 2 forks are made of 6 ounces of steel, and therefore 1 fork requires 3 ounces of steel.

There are other avenues to explore, such as starting with the fact that 3 forks and 3 spoons require 12 ounces of steel.

b. How many ounces of steel go into 1 spoon?

Solution: Similar explorations to part a can solve this problem. Or the student can use the result of part a: If:
3 forks and 1 spoon weigh 10 ounces, and

1 fork weighs 3 ounces, then
3 forks and 1 spoon weigh 9 ounces plus the weight of 1 spoon.
In other words:
9 ounces + weight of spoon = 10 ounces

And then you can see that the weight of the spoon is 1 ounce.

It helps students a lot to write down their chain of reasoning clearly. It's important for them to go over their solutions with at least one other person, or better still, present their solution to the club.

4-6. Race To The Middle

John and Paul go toward one another in a straight line. John walks at 2 miles per hour. Paul jogs at 4 miles per hour. They meet in 30 minutes. How far apart were they when they started?

Solution: In 30 minutes, John travels 1 mile. Paul travels 2 miles. They were 3 miles apart.

4-7. Dog Supplies

Sally has 4 dogs. She has enough dog food to last 30 days. She agrees to take care of 2 more dogs while her friends go on vacation. Assuming each dog eats the same amount of food each day, how many days will Sally's dog food last?

Solution: This is another job/rate/time problem in disguise. The job is to eat all the food, but the rate of eating goes up when more dogs are eating. In this case, the job is the same no matter how many dogs there are.

Each dog can eat all the food in 30x4 = 120 days.
Each dog eats 1/120 th of the dog food each day.
The rate for four dogs to eat all the food is 4x1/120 = 1/30 jobs/day.
And 4 dogs would be done in 30 days.

The rate for six dogs to eat all the food is 6x1/120 = 1/20 jobs/day.
6 dogs would be done in 20 days.

There are many good starting points for this problem. Another option: a student might realize that each dog eats ¼ of the food in 30 days. If there are 6 dogs, each dog only gets 1/6 of the food, so it won't last as long. The question becomes, how many days does it take a dog to each 1/6 of the food.

A student who is good at fractions might see that if a dog eats ¼ of the food in 30 days, the dog will eat 1/12 of the food, which is one-third of ¼ , in 10 days. In that case, the dog will eat 1/6 of the food in 20 days, because 1/6 is 2/12. It's worthwhile to explore the fractions in this problem.

4-8. Number Trick

Here is a number trick: Choose any positive whole number. Add 4 to it. Multiply the result by 2. Subtract your original number. Then subtract 5. Subtract the original number once more. Do this trick a few times with different numbers. You end up with a certain number every time.

a. What is the number you end up with?

Solution: 3

b. Explain why the trick works.

Solution: In the course of performing the trick the original number is added, doubled, and then subtracted twice. Therefore it vanishes from the calculation. All that is left is 4 multiplied by 2, followed by subtracting 5, which is always 3.

c. Does the mystery number have to be whole for the trick to work? Does it have to be positive?

Solution: No, and no. This trick works for any real number. If the chosen number is X, the trick is the following expression:
(X + 4)*2 – X – 5 – X which is 2X + 8 – 2X – 5 = 8 – 5 = 3.

(Where * means multiplication.)

4-9. Picking Peaches

An adult takes 3 hours to pick all the peaches from a tree. A child takes 7 hours to pick the same tree. How long would it take the adult and child to pick all the peaches from this tree, if they work together?

Give your answer in hours and minutes.

Solution: Another job/rate/time problem. The adult rate is 1/3 tree/hour. The child rate is 1/7 tree per hour. Together, if they finish one job in an amount of time T:
Rate x Time = Job

$\left(\frac{1}{3}+\frac{1}{7}\right)\times Time = 1\ Job$ Adding fractions: $\frac{10}{21}\times Time = 1\ Job$

which means that the Time must be 21/10 hours, which is 2 and 1/10th hours, or 2 hours, 6 minutes.

4-10. Filling A Draining Tub

A tub takes 4 minutes to fill, 5 minutes to drain. How long does it take for the tub to fill if the tap is running and the drain is open?
Professor Bear says: Don't do this at home.

Solution: The difference about this problem is that there is a positive rate for the tap: ¼ tub/minute, and a negative rate for the drain: -1/5 tub/minute
Rate x Time = Job

$\left(\frac{1}{4}-\frac{1}{5}\right)\times Time = 1\ Job$ Adding fractions: $\frac{1}{20}\times Time = 1\ Job$

which means that the Time must be 20 minutes.

How To Deal With Related Number Problems

The following are only suggestions:

Learn the vocabulary:

- Distance
- Rate
- Time
- Job
- Variable
- Equation
- Balance

Look for patterns such as:

- Sequence of actions (then work backwards)
- Problem involves distance, rate, and time (use the distance rate time relationship)
- Problem involves a job, rate, and time (use the job rate time relationship)

There might be surprises such as:

- A rate can be negative
- A problem might ask you to convert units (change seconds to hours, or centimeters per second to kilometers per hour, for example)

5. Statistics

Why Statistics

Here are Professor Bear's reasons to for the fifth chapter to be about Statistics:

- Statistics is a broad subject, encompassing counting, discrete math, and continuous math.
- Statistics has strong ties to applied mathematics, the mathematics of real-world data. This makes it an interesting counterpoint to pure math subjects. It shows how some pure math can be used in an applied situation.
- Statistics is a deep subject. It's impossible to get perfect data, and there are many interesting puzzles that have to do with how you get meaningful results from imperfect data.
- Statistics is used in experimental design and as such is part of *every* science. Statistics is used in the analysis of all measurements.

Underpinnings of Statistics

To understand this material you need some experience with arithmetic and be able to do one and two digit addition, subtraction, multiplication and division. You need to be able to read a graph.

It's also helpful but not necessary to know fractions.

Outcomes

After doing this chapter the student will know some statistics terminology and start to understand what the subject is about. They'll have some experience of the idea of experimental design.

Statistics: Notes and Vocabulary

Data is information you collect by taking measurements or counts.
The word "data" is plural and refers to many measurements or counts.

Some examples of data sets:

- The heights of all students in a particular classroom.
- The number of migrating birds that rest on a lake each day in the autumn.
- The temperature outside of your house measured every hour from 8:00 AM to 8:00 PM.

A **statistic** is a number or value calculated or derived from a data set. Here are some examples:

Data Set	Description of Data	Statistic
8, 9, 9, 9, 10	Ages of kids in a math club	**sample size**: 5 (there are 5 data points in the data set)
8, 9, 9, 9, 10	Ages of kids in a math club	**mean**: $\frac{8+9+9+9+10}{5}$ = 45/5 = 9 = average age
52, 53, 58, 60, 61	Heights of kids in math club, in inches	**median**: 58 (data point in the middle)
0, 0, 1, 1, 1, 2, 3, 3, 3, 4	Number of siblings of kids in school play	**sample size**: 10 **median**: $\frac{1+2}{2}$ = 1 and ½ (if the sample size is even, the median is the average of the middle two values)
khakis, jeans, chinos, jeans, jeans	Type of trousers worn by kids in math club	**mode**: jeans (most frequently occurring data point)
goldfish, cat, cat, dog, dog	Pets owned by kids in math club	**mode**: cat **mode**: dog (data set has two modes)

Mean

The average of n numbers is the sum of the numbers divided by n.
The average is also called the **mean**.

What is the mean of 1,2, and 3? 2

What is the mean of 0, 1, and 2? 1

Do you think it's possible for the average of 5, 5, 6, 7, 7, 8, 9 to be more than 10? What about less than 6? (Try to answer without calculating, and explain.)

Solution: No. There are seven numbers in the data set. If the average were 10 or more, then the sum of the seven numbers would have to be 70 or more. But each of the seven numbers is less than 10, so their sum is less than 70. Similarly, if the average is 6 or less, the sum of the seven numbers would have to be 42 or less. But some numbers are bigger than 6, so the sum is more than 42. The average is between 10 and 6.

Median

If a data set consists of numbers, the median is the middle value.

In a list with an odd number of numbers, the median is the exact middle value. In the data set 3, 5, 7, 10, 11, the median is 7.
What is the median of: 1, 1, 1, 2, 55, 56, 99? 2
Notice that the median is *not* the half-way point between the maximum and minimum values. The half-way point between 1 and 99 is 50.
What's the median value of 100, 101, 199, 200, 201, 399, 900? 200

In a list with an even number of data points, the median is the average of the two middle values.

What is the median of 2, 3, 4, 6, 9, 18? 5

What's the mean of 2, 3, 4, 6, 9, 18? 7

If the numbers in the data set have distinct (unequal) values, the median divides the data set into two halves: a half below the median, and a half above the median. If the data values are not distinct, this does not happen. Compare the medians of:
1, 1, 1, 1, 1, 1, 1, 7, 8, 9: the median is: 1. Less than half the data is above 1.
1, 3, 4, 6, 7, 9, 11, 12, 17, 20: the median is 8. Half the data is below 8, half above 8.

Mode

The mode of a list of numbers is the most frequently occurring number in the list.
The mode of 0, 0 , 0, 1, 2, 3, 3 is 0.

What about the mode of: 3, 3, 4, 4, 5, 5? This has three modes: 3, 4, and 5.

Write down a mode question for a friend:

What is the mode of: ________________________________

Solutions: Problems

5-1. What's A Statistic?

In the left column are statistics. In the right column are data sets. Match each statistic with a suitable data set.

Statistics	Data Sets
A. One out of three data points is an apple.	The age, in years, of each human being alive on Earth right now. D
B. The mode is 10.	The favorite hat worn by jazz dancers in the late 1980s. G
C. The mean is greater than 35.	The amount of time each book was checked out from the Vancouver Public Library, in the year 2000. E
D. The data values are between 0 and 130.	The number of salmon that swim to the mouth of the Skeena River each summer. F
E. The median is 19 days.	The ages of fifth graders in New York during the month of October in each year from 1960 to 2012. B
F. The mean is about 5 million.	The favorite fruit of each astronaut who rode a NASA space shuttle. A
G. There are two modes: the fedora and the porkpie.	The average retirement age of police officers in Elvewellyn. C

5-2. Page Number Statistics

Paul read a book with 37 pages. The pages are numbered 1, 2, 3, ... , 35, 36, 37.

a. What is the median page in Paul's book?

Solution: 19, the middle page number. There are 18 pages before page 19, and 18 pages after page 19. (Count to make sure.)

b. How many digits are used to number the 37 pages in Paul's book? Hint: the number of digits used to number pages 1 through 12 is 15.

Solution: First nine single-digit pages: 9 digits
Pages 10 through 19: 20 digits
Pages 20 through 29: 20 digits
Pages 30 through 37: 16 digits.
Total number of digits: 9+ 20 + 20 + 16 = 65

c. Paul reads a second book that has page numbers that use up 77 digits. What is the page number of the median page in the second book?

Solution: The answer to part b can be used to make this problem easier. If this book has pages that use 77 digits, that is 77 – 65 = 12 more digits than the previous book, corresponding to 6 more pages. The number of pages is thus 37 + 6 = 43. The median page is page 22.

5-3. Related Means

The mean (average) of four numbers is 17. The largest number of the four is 23. What is the mean of the remaining three numbers?

Solution: If the mean of the four numbers is 17, the sum of the four numbers is 4 x 17 = 68.

If the largest number is 23, the sum of the remaining three numbers is 68 – 23 = 45.

The mean of the remaining three numbers is 45 divided by 3, which is 15.

5-4. Mouse Data

Ten children try to measure the weight of a wiggly baby mouse. They don't get the same measurements. One child measured the weight as 2 grams. Three children measured the weight as 4 grams. Here is the measured data:

Weight measurement of wiggly baby mouse	Number of kids who got that measurement
2 grams	4
4 grams	3
5 grams	2
6 grams	1

Note to Coach: There is a reason why the mouse is described as a wiggly baby. The students are encouraged to imagine the difficulty of catching a tiny mouse, and having it lie still on a scale long enough to get a good reading of its weight.

a. Sketch the frequency distribution bar graph for the weight data.

Solution:

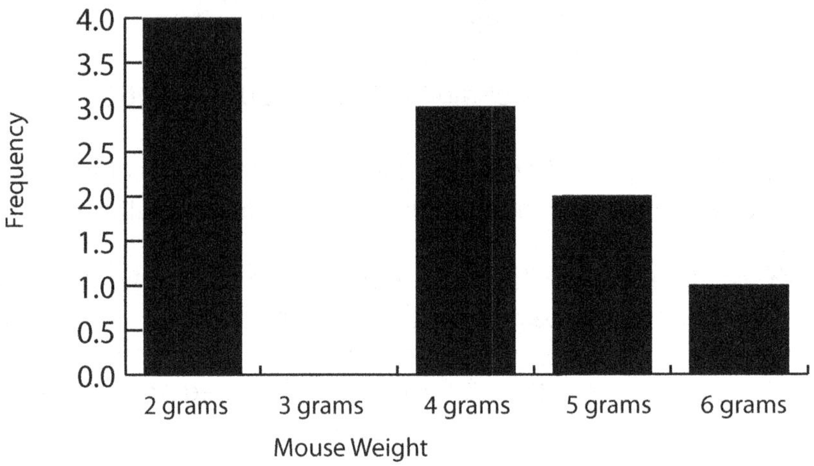

b. Calculate the mean, the median, and the mode of the weight data.

Solution: Number of data points: 4+3+2+1 = 10

Mean: $\frac{4\times2+0\times3+3\times4+2\times5+1\times6}{10}=\frac{36}{10}=3.6$

Median: data points are: 2, 2, 2, 2, 4, 4, ,4, 5, 5, 6. The two middle data points are 4. The mean of 4 and 4 is 4.

Mode: 2

c. What do you think is the best estimate of the true weight of the wiggly baby mouse? The mean? The median? The mode? Explain your conclusions.

Solution: There is no obvious best answer here. The data is somewhat skewed to the range from 2 to 4, so the mean, which is between 2 and 4, might be a best answer. A good thing to do would be to take more measurements and see if a clearer picture develops.

Note to Coach: Situations like this problem where there is not a clear answer provide room to discuss mathematical modeling. Measures of central tendency, (mean, media, and mode) are not "the Truth" about the data; they are only some statistics that we can calculate. It is up to us to figure out what those statistics mean, and whether we should pay attention to those statistics or not.

5-5. Five Mystery Numbers

A set of five positive whole numbers has a mean of 4, a median of 4, and a single mode of 6. What are the five numbers?

Solution: Let the five numbers be A, B, C, D, and E, in increasing order. If the median is 4, then C = 4. D and E must be more than 4, and the mode is 6, so D and E must both equal 6. If the mean is 4, then the sum of A+B+C+D+E must be 5 x 4 = 20.

Knowing that C=4 and D=6=E, we have
A + B + C + D + E = A +B + 4 + 6 + 6 = A + B + 16 = 20.
That means that A + B = 4.
A cannot equal B, because there is only one mode, so A and B cannot be 2.
Therefore A must be 1 and B must be 3.

5-6. Family Reunion

Thirty people attend a family reunion. The average age is 17 years. There are 12 girls, 10 boys, and 8 adults. If the average age of the girls is 10 and the average age of the boys is 7, what is the average age of the adults?

Solution: If the average age of the 12 girls is 10, then the sum of the ages of the girls is 120, because 120 divided by 12 is 10.

If the average age of the 10 boys is 7, then the sum of the ages of the boys is 70, because 70 divided by 10 is 7.

The sum of all ages is 30 x 17 = 510.

Therefore the sum of the ages of the adults is 510 – 120 – 70 = 320.
The average age of the 8 adults is 320 divided by 8, which is 40.

5-7. Averages Of Subsets

In a set of seven whole numbers, the average of the first four numbers is 6, and the average of the last four numbers is 10. If the average of all seven numbers is 8, what is the number common to the first four numbers and the last four numbers?

Solution: Let the seven numbers be A, B, C, D, E, F and G in increasing order.
The average of A, B, C, and D is 6. This means that A+B+C+D = 24.
The average of D, E, F, and G is 10. This means that D+E+F+G=40.
If the average of all seven numbers is 8, then A+B+C+D+E+F+G = 56.

If we add the sum of the first four numbers with the sum of the second four numbers we get 24 + 40 = 64. The difference between 64 and 56 is D, because D is counted twice. Therefore D = 8.

5-8. Rainfall

Here are annual rainfall bar graphs for two different locations, A and B.

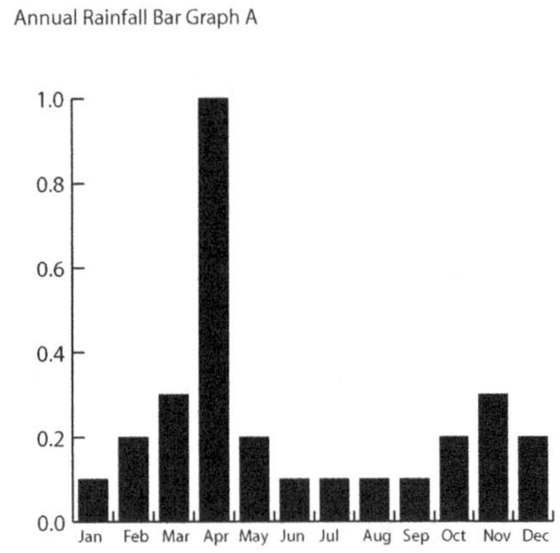

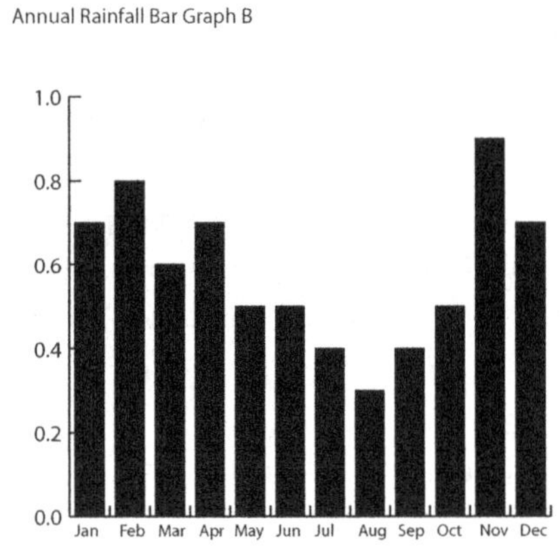

a. Which location has the higher average rainfall, A or B?

Solution: By inspection of the graph, B. It's possible to calculate, as well. A few calculations would show that the sum of the rainfall for the year for B is definitely larger than for A.

b. Which location has the highest rainfall in a month (in any month of the year)?

Solution: A, in April.

c. Which location has the lowest rainfall in a month?

Solution: A, in January, June, July, August and September.

d. Which location, A or B, is most likely to be a desert?

Solution: A. It has less rain and higher extremes.

e. Which location has the most **even** distribution of rainfall? Use numbers to justify your answer.

Solution: A pure even distribution is easy to define. Its bar graph would consist of bars that all have the same height—the same rainfall every month. What is harder to define is what makes one distribution "more even" than the other? In statistical analysis, that question has an answer but it's beyond the scope of this book. However, for elementary school students, a good start is to compare the ranges of the two graphs. A perfectly even distribution has a data range of zero: the maximum minus the minimum is zero.

At first glance, graph B has the smaller range. The rainfall for A ranges between 1.0 in a month and 0.1 in a month—a spread of 0.9. The range for B is between 0.9 in November and 0.3 in August—a spread of 0.6.

However, it's worth noting that if you remove the heavy rainfall month of April from graph A, which is an outlier, the data range for A is from 0.1 to 0.3, a spread of only 0.2. Perhaps the stronger case for the "more even" distribution is graph A.

You could ask your students to brainstorm on how to define "more even" and "less even" in these kinds of circumstances. In statistical analysis the variance is used in this way. The variance is a measure of the deviation of the data points from the mean.

5-9. Truth And Statistics

Compare and contrast statement A and statement B:

A Out of 100 dentists surveyed, 80 dentists said they preferred Whizzy-Floss to all other brands of dental floss.

B 4 out of 5 dentists recommend Whizzy-Floss.

Note to Coach: You might want to explain to your students that the phrase "Compare and contrast" means: "specify what is the same, and what is different about ..."

Solution: A and B are both statistics. They both involve the fraction 4/5, or 80%. A is a more specific statement that only refers to 100 dentists surveyed.

Be is a more general statement than A, and seems to be based on all dentists. There is some ambiguity in B. Does it mean that if you took a random sample of any 5 dentists anywhere in the universe, 4 of them would recommend Whizzy Floss? Or does it mean that out of all the dentists in the universe, 80% of them recommend Whizzy Floss (which would mean that there might be a group of 5 dentists that don't recommend Whizzy Floss).

Statement A does not have the ambiguity of statement B.

5-10. Statistics Problem 10

Here is a small data set:

7, 7, 8, 8, 10

a. Calculate the mean, median, and mode.

Solution: Mean: 40/5 = 8

Median: 8

Mode: two modes, 7 and 8

b. How much would you have to change the first data point to lower the mean by 1 unit? What would happen to the median and mode if you lower the first data point?

Solution: To lower the mean by 1 point, the sum of the numbers has to be 35 instead of 40. So the first data point would be 2 instead of 7. The median would be the same, and the mode would be only 8.

c. How much would you have to change the last data point to raise the mean by 1 unit? What would happen to the median and mode if you raise the last data point?

Solution: By the same reasoning, the 10 would have to be raised to 15 to raise the mean by 1 unit. The median would be unchanged, and the mode would be only 7.

d. Do you see a pattern to the answers of (b) and (c)? What if, instead of five data points in the data set, there are 100 data points—how much would the data values have to change to shift the mean up by one unit?

Solution: The pattern is that for the mean to go up or down 1 unit, one data point would have to change by 5 units. If there were 100 data points, one data point would have to change by 100 units to effect a 1-unit change in the mean.

How To Deal With Statistics Problems

The following are only suggestions:

Learn the vocabulary:

- Data
- Data set
- Statistic
- Mean
- Average
- Median
- Mode

Look for patterns such as:

- If the average of 5 numbers is 6, then the sum of the 5 numbers is 30.
- If there are 4 numbers and the average goes up by 1 point, the numbers must have increased by 4.

There might be surprises such as:

- The mean of a set of whole numbers might not be a whole number.
- Means, medians, and modes might be very different from one another.
- Some data sets have more than one mode.
- The median is not necessarily the halfway point between the maximum and minimum of a data set.

6. Logic

Why Logic

Here are Professor Bear's reasons to for the sixth chapter to be about Logic:

- Logic is a muscle that grows as you do more math and more math.
- Logic is at a crossroads between numbers and language.
- Logic is at a crossroads between math and philosophy.
- Basic logic is learned at a very young age. Students use a lot of basic logic without being aware of it. There is a pitfall of believing that correct answers are "obvious" or only require "common sense". This is a dangerous pitfall, as Shakespeare famously pointed out: all is not what it seems. Logic problems are a good place to learn the value of careful thinking.
- Learning about logic can feel heavy, but the point is not to be heavy and give names to obvious things. The point is to learn about subtleties in reasoning and use them to advantage.
- Another point is to learn to argue and disagree like civilized beings.
- Studying logic means studying types of statements and types of arguments. This enables the student to recognize: Oh, this is a certain kind of problem and I know how the solution has to go.
- Like statistics, logic helps us with imperfect information. We might not know the whole story, but we have clues. How do we get the full story from the partial clues?

Underpinnings of Logic

To understand this material you need some experience with arithmetic and be able to do one and two digit addition, subtraction, multiplication and division. You need to be able to read a graph.

It's also helpful but not necessary to know fractions. It helps to read well and carefully.

Outcomes

After doing this chapter students will be able to recognize statements, implications, and converses. Students will be able to do some logic puzzles. They might absorb some techniques for civil discussion and logical argument.

Logic: Notes and Vocabulary

Logic is the study of what is true and what makes sense. A frequent question in logic is: how do you assemble small truths into big truths? In other words, can a few clues (assorted little truths) lead to the solution of a mystery?

In math, a basic building block is number. We often work with numbers.
In logic, a basic building block is a **statement**. A statement is a sentence that can be true, or false, but not both.

Statements:

Even numbers are divisible by two.
Even multiplies of five end in zero.
My brother lives on Mars.

Not statements:

Hey, Bob!
I am lying.

Clue Puzzles

There are many puzzles where you have a few true statements about something, and the goal of the puzzle is to figure out everything from a few true statements.

Example: Melissa, Susan, and Phil live in the same town. One of them is 10, one is 11, and one is 12. Phil is the same age as Susan's younger brother. Melissa is one year older than Phil. What are the ages of Melissa, Susan, and Phil?

A good way to solve this puzzle is by making a chart of all the possible combinations of names and ages. Then cross off combinations that are ruled out by the clues:

	Melissa	Susan	Phil
10			
11			
12			

One clue says: *Phil is the same age as Susan's younger brother*. This means that Phil is younger than Susan, and therefore Phil cannot be the oldest person. Also, Susan cannot be the youngest person. From this clue alone we can cross off two possibilities:

	Melissa	Susan	Phil
10		X	
11			
12			X

The next clue is: *Melissa is one year older than Phil.* This means that Melissa cannot be the youngest person. Here is the updated chart:

	Melissa	Susan	Phil
10	X	X	
11			
12			X

Because one of the three people must be 10, the 10 year old must be Phil. Here is the update of the chart:

	Melissa	Susan	Phil
10	X	X	√
11			X
12			X

Since Melissa is one year older than Phil she must be 11, and then Susan must be 12.

Another example of a clue puzzle is a Number Place puzzle, also known as Sudoku. Here is a simple Number Place puzzle:

2		4	
3		2	
	3		

The rules of this puzzle are that each row, each column, and each outlined two-by-two square must contain the digits 1, 2, 3, and 4.

Because, for example, the upper left square has to contain all four numbers, the two empty spots must contain 1 and 4. Because 4 is in the top row already, the lower spot must contain the 4:

2	1	4	
3	4	2	
	3		

You can now proceed to fill in the rest of the spots. In the first row, the missing digit is 3; in the second row, it is 1, and so on.

You can see how these puzzles are chains of logically related true statements:
Statement: There are a 2, 1, and 4 in the upper row.
Statement: Each row must have all four digits, 1, 2, 3, and 4.
Statement: Therefore, the upper row must have a 3 in the right-most spot.

Put another way, solving the puzzle means making a chain of logically related true statements. The chart and the Number Place grid are ways to organize the clues so that the solution becomes easier to see.

Cipher Puzzles

A cipher is a translation of letters into numbers or numbers into letters. Using logic from arithmetic, you can solve simple cipher puzzles.

Example: The letters U, R, and X represent three different single positive digits. Find out what the digits are if they obey this sum:

$$\begin{array}{r} U \\ +R \\ \hline RX \end{array}$$

Solution: Because U and R are single digits, the sum RX has to be less than 20. That means that R has to be 1. Now we can translate part of the sum:

$$\begin{array}{r} U \\ +1 \\ \hline 1X \end{array}$$

This means that U + 1 = 1X
The only one-digit number that becomes a two-digit number if you add 1 to it is 9 (any other digit is too small). So U must be 9. And then X must be zero (0).

Venn Diagrams

Here is an example of a logic problem:

At Smithson School, there are 50 fifth graders. They all wear some form of school clothing: a hat or a vest. 25 fifth graders wear school hats. 35 fifth graders wear school vests. How many fifth graders wear both hat and vest?

Solution 1: the total number of students is 50. 25 + 35 = 60. So there must be some students who wear both hats and vests. 60 - 50 = 10, so there must be 10 students who wear both hats and vests.

Solution 2: Venn diagram.

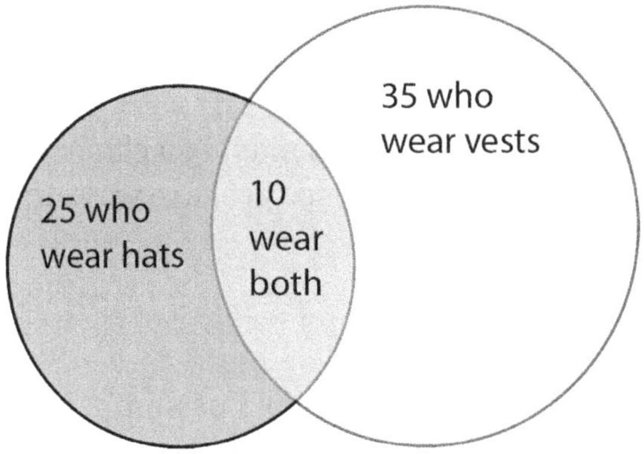

The Venn diagram above shows a circle for the 25 who wear hats, and an overlapping circle for the 35 who wear vests. The overlap of the two circles represents students who wear both hats and vests.

You can see that the group of 25 students is made up of those who wear hats but not vests plus those who wear both hats and vests. You can write an equation:

25 = hats-only fifth graders + fifth graders who wear both

Write a similar equation for the group of 35 students who wear vests:

Solution: 35 = vest-only fifth graders + fifth graders who wear both

Write a similar equation for the total group of 50 students:

Solution: 50 = (hats-only fifth graders) + fifth graders who wear both + (vest-only fifth graders).
This translates to: 50 = (25 – both) + both + (35 - both) = 60 – both and so the number that wear both must be 10.

Debunking

It is often harder to prove that something is true than to prove that something is false. All it takes to prove a statement is false is one example of falseness—a **counterexample**.

It is worthwhile to practice the ability of coming up with good counterexamples. Here are a couple of outlandish statements. Come up with counterexamples:

Note to Coach: The following counterexamples are only suggestions. If you do any of these as a group, it's useful to show several different valid counterexamples.

Statement: If a number is divisible by 2, it is also divisible by 62.

Counterexample: **The number 4. 4 is divisible by 2, but not by 62.**

Statement: If a number N is divisible by 7, then N+1 is divisible by 8.
(Notice that this statement is true if N=7.)

Counter example: **The number 14. 14 is divisible by 7, but 14 + 1 = 15 is not divisible by 8.**

Here are some more statements that are false. Come up with counterexamples:

Statement: There are no prime numbers between 30 and 39.

Counterexample: **The number 37 is between 30 and 39 and is prime.**

Statement: If you measure a rectangle's area in square inches and its perimeter in inches, it is not possible for the perimeter to be a larger value than its area.

Counterexample: **A rectangle of width 1 and length 3 has area 3 square units, and perimeter 8 units.**

Statement: If an apple weighs the same as a banana and a tangerine, then the apple weighs the same as two bananas.

Counterexample: **If the banana weighs 2 pounds and the tangerine weighs 1 pound, the apple weighs 3 pounds. Two bananas weighs 4 pounds, which is not the same as the weight of the apple.**

6-1. Adding Mystery Digits

Three digits, A, B, and C obey this sum:

$$\begin{array}{r} AAA \\ +BBB \\ \hline AAAC \end{array}$$

What are A, B, and C?

Solution: Digit A must be 1, because the sum of A and B must be less than 20.

This means the sum looks like:

$$\begin{array}{r} 111 \\ +BBB \\ \hline 111C \end{array}$$

The only digit that B can be is 9, because 1+B has to be at least 10.

Now the sum looks like:

$$\begin{array}{r} 111 \\ +\ 999 \\ \hline 111C \end{array}$$

And C has to be zero. The solution is that A, B, and C are 1, 9, and 0.

6-2. Multiplying Mystery Digits

Two digits, A and B, obey this product:

$$\begin{array}{r} ABBA \\ \times\ \ AA \\ \hline AABAA \end{array}$$

What are the digits A and B?

Solution: One place to start: if AxA has a ones digit of A, then A is either 1, 5, or 6.

The next observation is that we have a 4-digit number multiplied by a 2-digit number to get a 5-digit number. The smallest 6-digit number is 100,000, so that means that AABAA is 99,000 or less.

If A=5, then the product becomes:

$$\begin{array}{r} 5BB5 \\ \times\ \ 55 \\ \hline 55B55 \end{array}$$

but this is impossible because 5000 x 50 is 250,000 which is much larger than 55B55 could be. This means that A cannot be 5 or 6, and A must be 1.

Now the product becomes:

$$\begin{array}{r} 1BB1 \\ \times\ \ 11 \\ \hline 1BB1 \\ 1BB10 \\ \hline 11B11 \end{array}$$

This means that B = 0.

6-3. Number Place Puzzle

Fill in the Number Place puzzle. The rules are: each row contains the digits 1, 2, 3, 4; each column contains 1, 2, 3, 4; and each outlined 2x2 square contains 1, 2, 3, 4.

			1
	4		2
	3		

Solution: The solution of this puzzle involves logical steps:
In the upper right 2x2 box, there is a 1 and 2. The remaining boxes must have 3 and 4. The 2 occupies a row with 4, and so the box beside the 2 has a 3. The box beside the 1 has a 4.
Then the second row becomes 1 4 3 2. And so on.
Final entries:

3	2	4	1
1	4	3	2
2	3	1	4
4	1	2	3

6-4. Cross Sum Puzzle

Here is a Cross Sum puzzle. You fill it in with numbers that add up to the white numbers on the left, if the numbers are along a row, or add up to the white number above, if the numbers are along a column. No digits can be repeated.

Examples:

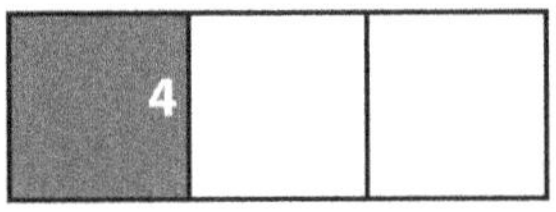

The empty squares above can be filled with 1 and 3, 3 and 1, but not 2 and 2.

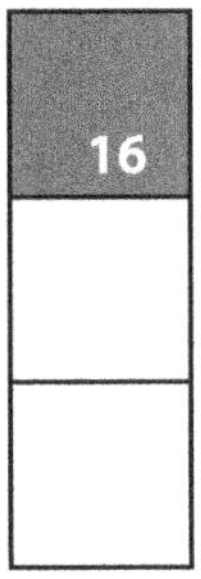

The empty squares at left can be filled with 7 and 9, 9 and 7, but not 8 and 8.

Some useful sums:		
	1 + 2 = 3	1 + 3 = 4
	1 + 2 + 3 = 6	1 + 2 + 4 = 7
	1 + 2 + 3 + 4 = 10	1 + 2 + 3 + 4 + 5 = 15
	7 + 9 = 16	8 + 9 = 17

What is special about all of these particular sums? (Compare with the possible ways to sum up to 5, or 8, or 9).

Solution: These particular sums are "unique" (disregarding the order of the terms). In the case of 16, for example, no other two unequal numbers

besides 7 and 9 sum to 16. The number 10 is very different: 1+9=10, 2+8=10, 3+7=10, 4+6=10.
Final entries:

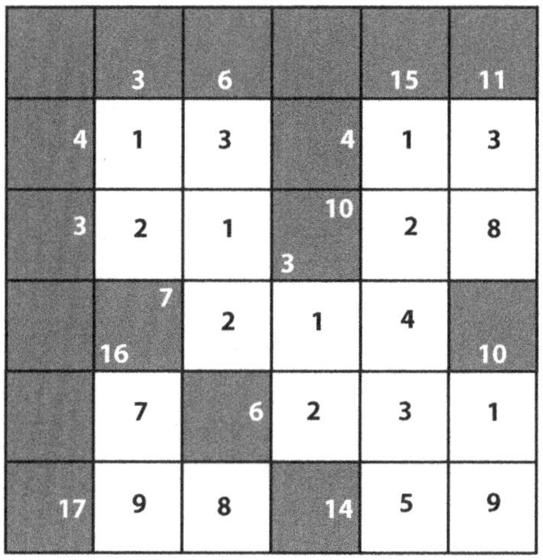

6-5. Who Made What Jam?

Penny, Lucy, Max, and Carl each made jam. The jams they made, in no particular order, are strawberry, raspberry, peach, and fig.

Penny and Carl made berry jams.
Max is allergic to peaches.
Lucy waved hello to Carl at the raspberry farm.

What type of jam did each person make?

Solution: This problem is perfect for the checkbox strategy. The first clue, Penny and Carl made berry jams, excludes peach and fig for Penny and Carl. This is shown below.

	Penny	Lucy	Max	Carl
Strawberry				
Raspberry				
Peach	X			X
Fig	X			X

The second clue excludes peach from Max, which means that Lucy made peach jam. Because Lucy did not make fig jam, Max made fig jam. And so on.
Solution: Max, fig. Lucy, peach. Carl, raspberry. Penny, strawberry.

6-6. Who Laid Green Eggs?

Mera, Ciela, and Terra are three chickens. They each lay eggs of a particular color. One day they laid a total of 10 eggs. Some eggs are white, some are brown, and some are green.

Ciela laid the most eggs by 1.
There are equal numbers of white eggs and green eggs.
Mera's eggs are green.

How many eggs of each color did each chicken produce?

Solution: A checkbox table can help determine the color of each chicken's eggs.

If the total number of eggs is 10, and if there are equal numbers of white and green, then the numbers of eggs are either: 1, 1, 8; or 2, 2, 6; or 3, 3, 4; or 4, 4, 2. Because Ciela laid the most eggs by 1, the only combination that works is 3, 3, 4. Ciela's eggs can't be white or green, so she must have laid 4 brown eggs. Mera's eggs are green, so she laid 3 green eggs. Terra laid 3 white eggs.

6-7. Implications And Converses

An **implication** is a special kind of statement. It is an "If ... then ..." statement.
Examples of implications:
If Willie sings, then Willie is happy.
If Horace makes a cake, then the kitchen needs cleaning.
If it rains, then Appleton is happy.

The converse of an implication is a related "If ... then ..." statement where the first and second parts are switched. Here are the converses of the example implications above:
If Willie is happy, then Willie sings.
If the kitchen needs cleaning, then Horace makes a cake.
If Appleton is happy, then it rains.

You can see from these examples that implications might be quite different from their converses. And if the implication is true, the converse might be false. Here are some mathematical implications. State whether they are true or false, and write down the converse statement. State whether the converse is true or false.

Implication	True or False?	Converse	Converse True or False?
If a number is even, then the ones digit of the number is either 0, 2, 4, 6, or 8.	**True if whole**	**If the ones digit of a whole number is 0, 2, 4, 6, or 8, then the number is even.**	**True**
If a number is prime, then the number is not divisible by 4.	**True**	**If a number is not divisible by 4, then it is prime.**	**False**
If the remainder of a number is 1 when you divide it by 5, then the number has a ones digit of 1.	**False**	**If a whole number has a ones digit of 1, then it has a remainder of 1 when you divide it by 5.**	**True**

6-8. Debunking Practice

Here are some statements. Determine if they are true or false. If a statement is false, give a counterexample.

Statement	True or False?	Counterexample if False
Any multiple of 10 is divisible by 5.	**True**	
Any multiple of 5 is divisible by 10.	**False**	**15 = 3x 5 but is not divisible by 10.**
All primes except 2 are odd.	**True**	
The largest number is googol.	**False**	**googol+1 is larger than googol.**
If good numbers are divisible by 3, and bad numbers are divisible by 17, then a number cannot be both good and bad.	**False**	**51 = 3x17 is divisible by both 3 and 17, so it is both good and bad.**
If X is a number that is 5 less than another number Y, then X is more than half of Y.	**False**	**X = 3 is 5 less than Y=8, but half of Y is 4 which is more than X. Notice that the statement is true if X > 5.**

6-9. Class Pets

A teacher recorded some data about the pets owned by children in his classroom.
15 children have dogs.
18 children have cats.
8 children have fish but no dogs and no hamsters.
6 children have hamsters but no cats and no fish.
3 children have hamsters and dogs.
7 children have both dogs and cats.
4 children have fish and cats.

How many children are there in the class?

Note to Coach: This is a classic Venn diagram problem. The wrinkle is that not all of the circles touch each other. A good question to ask is: if 8 children have fish but no dogs and no hamsters, what other pets could they have?

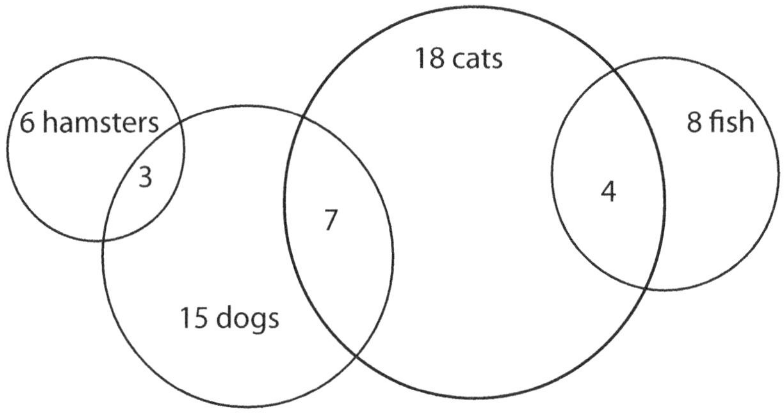

Other questions:
How many children have only dogs, no other pets? 5
How many children have only cats, no other pets? 7

Solution: There are 33 children in the class: 3 with hamsters only, 3 with hamsters and dogs, 5 with dogs only, 7 with dogs and cats, 7 with cats only, 4 with cats and fish, and 4 with fish only: 3 + 3 + 5 + 7 + 7 + 4 + 4 = 33. There are numerous ways to do the accounting. They could for example add 6 + 15 + 18 + 8 and subtract all overlaps (3 + 7 + 4). 47 – 14 = 33.

6-10. Twin Primes

Twin primes are consecutive prime numbers that differ by 2. The numbers 3 and 5 are twin primes. Another set of twins is 11 and 13.

a. Find all of the twin primes less than 50.

Solution: This is a good place to inform students about the Sieve of Eratosthenes, an ancient method of finding primes. The method is simple: array the whole numbers in a grid and start by crossing off multiples of 2. Then multiples of 3, multiples of 5, and so on. (You won't have to cross off multiples of 4 because they are already crossed off.

Twin primes: 3 & 5, 5 & 7, 11 & 13, 17 & 19, 29 & 31, 41 & 43

b. 3, 5, and 7 are "triplet" primes, three consecutive primes that increase by 2. Are there any other sets of triplet primes? Explain why or why not.

Solution: Suppose you have twin primes bigger than 3 and 5. Then you have three consecutive numbers:
Prime One, Middle, Prime Two. In any set of three consecutive numbers, one of them is a multiple of 3.

Think about it:
If the first number is not a multiple of three, its remainder when divided by 3 is either 2 or 1. Then, the second number, when divided by 3, has a remainder of 1 or 0. If the remainder is 0, it is a multiple of 3. If the remainder of the second number is 1, then the remainder of the third number is 0, and the third number would be a multiple of 3.

Going back to twin primes, the Middle number has to be a multiple of 3, because Prime One and Prime Two cannot be multiples of 3.

If there is a triplet of primes, you'd have five consecutive numbers:

Prime One, First Middle, Prime Two, Second Middle, Prime Three

We know for sure that First Middle is a multiple of 3. But Prime Three is 3 units bigger than First Middle, which means that Prime Three must also be a multiple of 3. Oh oh. What this means is that Prime Three, except in the special case of 3, 5, and 7, never exists.

This is an example of a very math-y kind of problem. Students can do it but might need help thinking through the steps and writing down their solutions in a coherent way.

c. Can you explain why the number between a pair of twin primes is always divisible by 6? (With the one exception of 4.)

Solution: All the hard work for this problem is already done in part b. We know that any time there are twin primes, the middle number is divisible by 3. We also know that the middle number is divisibly by 2, because the two surrounding prime numbers must be odd. Because the middle number is divisibly by both 3 and 2, it is divisible by 6.

How To Deal With Logic Problems

The following are only suggestions:

Learn the vocabulary:

- Statement
- Counterexample
- Implication
- Converse
- Venn diagram

Look for patterns such as:

- A word problem that consists of a set of clues might turn out to be solvable by a clue checkbox.
- A statement that feels false can often be proved false with one counterexample.
- In a mystery digit problem, consider the place value.

There might be surprises such as:

- In a Venn diagram of a situation, not all circles necessarily intersect. Some circles might not touch others, and some circles could stand alone.
- The converse of a true implication might be false.
- The converse of a false implication might be true.

7. Area And Perimeter

Why Area And Perimeter

Here are Professor Bear's reasons to for the seventh chapter to be about Area And Perimeter:

- The measurement of area, a two-dimensional quantity, is a big leap from the measurement of length. In the context of area, a very long line has zero area.
- People have very variable amounts of experience with area. People who sew might have a lot of intuition coming from areas of cloth—they know what a square yard or meter looks like, or a half-square yard. Someone who bakes pizzas might too. And someone who does neither of these things might have very little intuition at all.
- A person can be taught in school that area is length times width, and have no idea how to estimate the area of her desk or the outer area of a soda can. It's very helpful to build up these intuitions.
- Area provides great examples of multiplication, division, and factor problems.
- The perimeter:area ratio is a far-reaching concept that students will see again and again, in biology, physics, architecture, mathematics. It's analogous to the surface:volume ratio.
- The perimeter:area ratio has two aspects to it: one, its variability. A fixed perimeter can correspond to many different areas. A fixed area can correspond to many different perimeters.
- The second important aspect of the perimeter:area ratio is its relationship to shape. The squarer or rounder the shape is, the lower the ratio is. In more advanced math we prove that there is a minimum value of the ratio, for the circle. (The circle is the shape with the least perimeter per enclosed area.)
- Areas of triangles and circles are included in the chapter, because there is more to area than the product of length and width.

Underpinnings of Area And Perimeter

To understand this material you need some experience with arithmetic and be able to do one and two digit addition, subtraction, multiplication and division. It helps to know about some basic geometrical shapes: squares, rectangles, circles, triangles.

When working with circles, it is good to be comfortable with non-whole numbers.

Outcomes

After doing this chapter students will have more experience estimating and measuring areas, solving arithmetic problems with areas, and understanding the relationship between perimeter and area.

Area and Perimeter: Notes and Vocabulary

Length is the one-dimensional size of something. You can measure length with a ruler.

Area is the two-dimensional size of something: how far does it spread out in two directions. People want to know areas of carpets, lawns, cloth, countries. Here are some questions to think about:

How would you measure the area of a tiled floor?	Discuss.
How would you measure the area of a house window?	Discuss.
How would you measure the area of a car windshield?	Discuss.
What might an "area ruler" look like?	Discuss.

One approach to measuring areas is to cover them with smaller units of a known area. The easiest area to measure is the area of a rectangle, because you measure the length and width of the rectangle with a ruler, and then multiply the length by the width to get the area.

Each of the rectangles below is labeled with its dimensions in Larks.

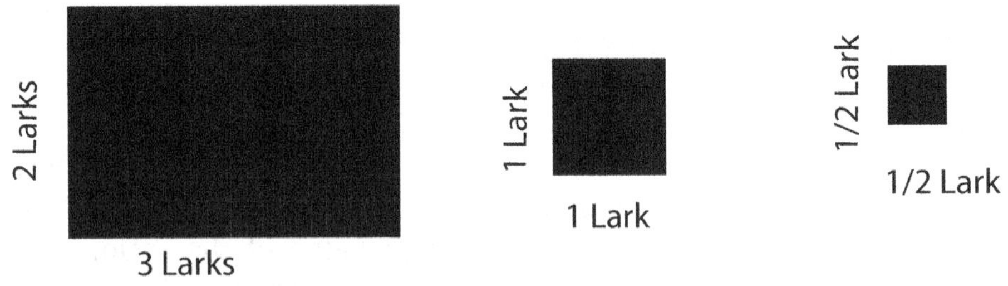

What is the area of each rectangle, in square Larks?

Solution: 6 square Larks, 1 square Lark, ¼ square Lark

Is the area "bigger" than the length and width? Why or why not?

Solution: This is not a well-posed question. See below.

You can see from the three rectangles above, sometimes the size of the area is larger than either length or width, sometimes it is the same, and sometimes it is smaller.

The question of whether the area is "bigger" than the length is an example of a badly posed question. Areas and lengths are different measurements. Notice that the area is measured in **square Larks**, and the length in **Larks**. You can't compare measurements taken in two different types of units.

Another thing to notice, however, is that because rectangular areas are products they have a strong connection to multiplication and factors. The length and width of a rectangle are factors of its area.

Many rectangle problems involve multiplication and factors, and many multiplication problems can be solved using rectangles.

A final observation: the quirks of multiplication apply to areas. Sometimes when you multiply two numbers, you get a number that is bigger than the original numbers. Sometimes the answer is much smaller. What is the area of a rectangle that measures 0.01 Lark by 0.001 Lark?

Solution: The dimensions can be rephrased as 1/100 Lark by 1/1000 Lark, which might be easier to multiply. The solution is then 1/100,000 square Lark, which is 0.00001 square Lark. This is a good time to look at different methods of multiplying decimals.

Perimeter is the distance around an area. It is a one-dimensional size that you can measure with a ruler.

What is the perimeter of each rectangle on the previous page?

Solution: 10 Larks, 4 Larks, 1 Lark.

Solutions: Problems

7-1. Square Tiles

Jonie has 24 square tiles, one-inch by one-inch. With the tiles she can make rectangles.

a. Whatever she makes, if she uses all of her tiles with no overlapping, the area is **24**.

b. How many rectangles of different dimensions can she make? The dimensions are the length and width of the rectangle. For example, she can make one rectangle that is 3 tiles wide and 8 tiles long. What are the others?

Solution: 1x24, 2x12, 3x8, 4x6

c. What is the largest perimeter rectangle she can make?

Solution: 1x24, perimeter = 50. A far-reaching insight is that long skinny shapes have larger perimeters than square shapes. This is the surface to volume ratio concept that is so influential in architecture and biology.

d. What is the smallest perimeter rectangle she can make?

Solution: 4x6, perimeter = 20. Again, this fuels the intuition that the closer the rectangle is to a square, the smaller the perimeter is. A house with a more square cross-section would have a smaller perimeter, a smaller surface area, and lose less heat through its walls in the winter.

Another important intuition to develop here is that area and perimeter are different things. The same amount of area could have infinitely many different perimeters—depending on the shape.

7-2. Folded Paper Part One

A square of paper of area 64 square cm is folded in half to make a rectangle. What is the perimeter of the rectangle?

It's a great idea to draw a diagram for this problem:

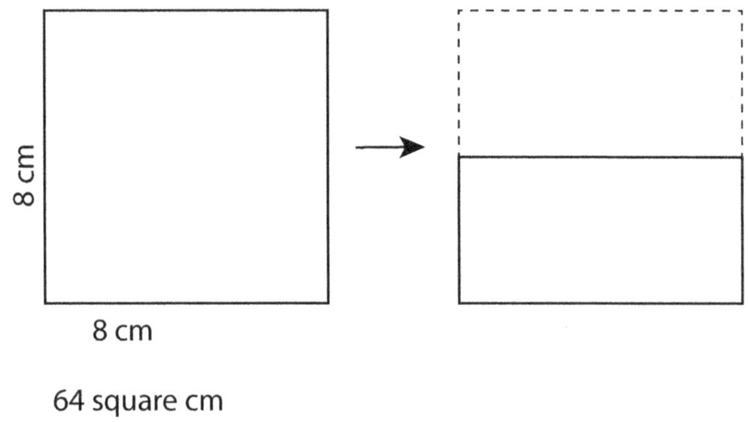

From the diagram, the dimensions of the new rectangle are easier to see.

Solution: The new rectangle is 4 cm x 8 cm, with perimeter 8cm + 16cm = 24 cm.

7-3. Folded Paper Part Two

A square piece of paper is folded in half to make a rectangle. The perimeter of the rectangle is 30 cm. What is the area of the square?

Solution: It really helps to do problem 7.2 first, which is easier. With a similar diagram, the student can see that if the side of the square is S cm, then the perimeter of the rectangle is S + S + ½ S + ½ S = 3S cm. And so S must be 10. (In the diagram for 7-2, replace 8 with S). The area of the square is 100 square cm.

7-4. Rectangular Tiles

Edgar has three rectangular tiles that are 1 inch wide and 2 inches long.

a. How many different ways can he arrange them in a rectangle of area 6 square inches? Assume that the tiles can be either vertical or horizontal when arranged.

Solution: The three different arrangements are these:

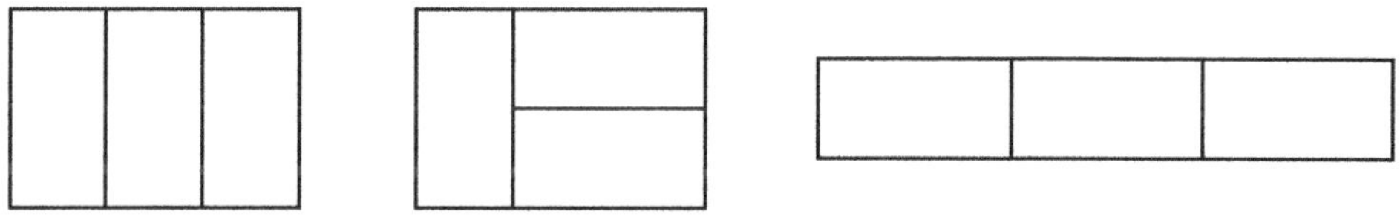

b. What perimeters are possible for Edgar's 6-square-inch rectangles?

Solution: There are two possible perimeters: 10 and 14. Refer to the diagram in part a.

c. Edgar's rectangles fall into different groups according to symmetry. Two rectangles are in the same symmetry group if it is possible to move one on top of the other and they match shape exactly. What are the different symmetry groups?

Solution: There are two outer shapes in the three arrangements of part a. One shape is a 2x3 rectangle, and the other is the 1x6 rectangle. These two shapes represent the two symmetry groups that Edgar's rectangles could belong to.

7-5. Changing Areas

Ralph has 9 tiles of area one square inch. He arranges them in a 3x3 square of perimeter 12 inches.

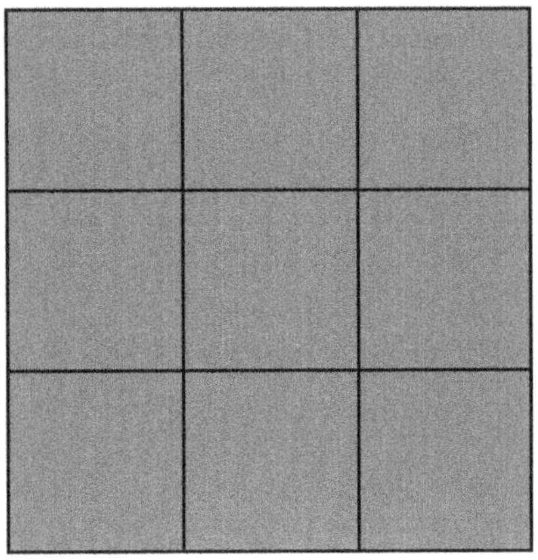

Ralph removes two tiles and gets the area below.

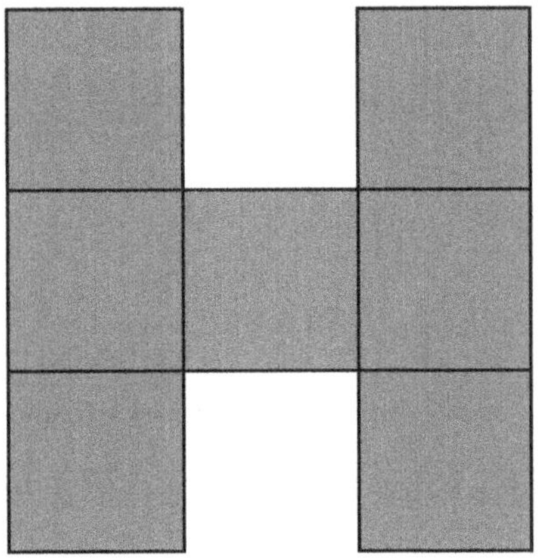

a. What is the area of the second figure?

Solution: Count the squares in the second figure. The area is 7 square inches.

b. What is the perimeter of the second figure?

Solution: Count the outer sides in the second figure. The perimeter is 16 inches.

Observation: Notice that going from the square to the "H" shape, the area gets smaller (9 to 7) but the perimeter gets larger (from 12 to 16).

7-6. Circumference

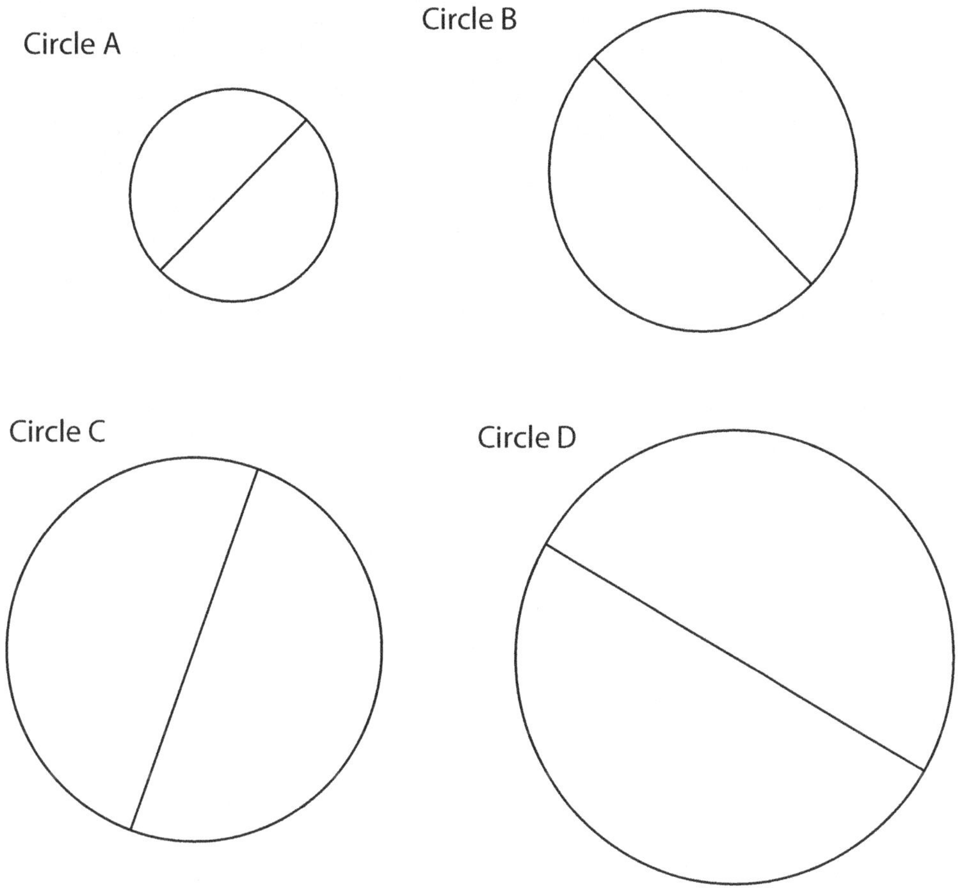

Measure the perimeters and diameters of Circle A, Circle B, Circle C, and Circle D. The perimeter of a circle is also known as the circumference. The diameter of a circle is the widest distance across the circle. Record your data in the table below. Hint: use string to measure the circumference.

Circle	Circumference	Diameter	Circumference ÷ Diameter
A			
B			
C			
D			

What do you notice about the circumference ÷ diameter?

Note to Coach: The point of this problem is to get experience with π and especially with the fundamental definition of π (Pi), which is the ratio of circumference to diameter. It's interesting to ask students, before doing this problem, what they might already know about π. Often students will have heard that the area of a circle involves π. It is quite difficult with the circles on the page to get a very accurate value of π. Typically students get values of around 3.2. Here is π to 100 decimal places:

3.14159 26535 89793 23846 26433 83279 50288 41971 69399 37510 58209 74944 59230 78164 06286 20899 86280 34825 34211 70680

7-7. Circles Vs. Squares

Imagine a square. You can draw two circles related to the square:

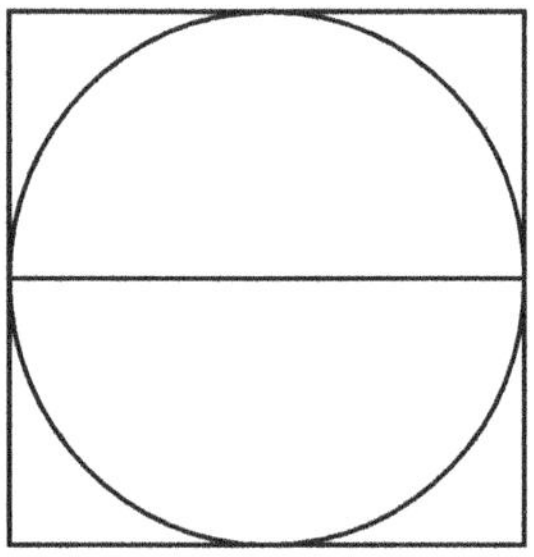

Diameter = side length
Circle is smaller

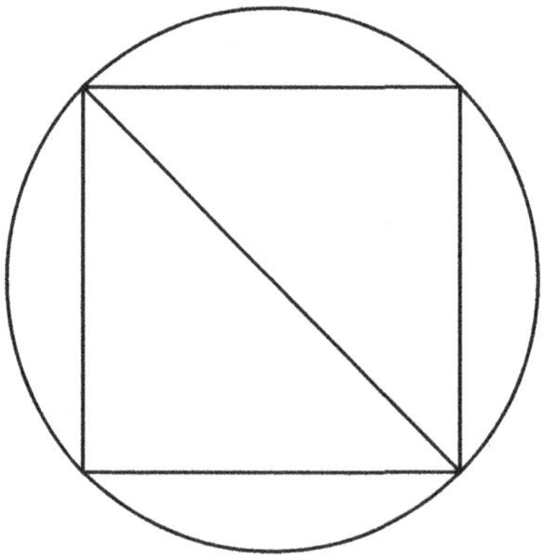

Diameter = diagonal length
Circle is bigger

In the first case, the circle is smaller (in area) than the square. In the second case, the circle is larger in area than the square. There is a famous ancient problem called "Squaring the Circle" where mathematicians looked for an easy way to find a square and a circle with the same area. They looked for a relationship between the length of the side of the square and the diameter of a same-size circle. This problem turned out to be more difficult than anyone expected.

Note to Coach: The above paragraph simplifies the explanation of the ancient problem of Squaring the Circle. More accurately, Squaring the Circle means constructing a square with the same area as that of a given circle using only a straightedge and compass. It's easy to google Square The Circle and find lots of history and background.

In the previous problem, you measured the circumference and diameters of four circles. It turns out that for *every* circle, the circumference ÷ diameter is equal to the same number, a special number a little larger than 3 called π.

Using π, we have easy formulas for the circumference (perimeter) and area of circles. The circumference is π x Diameter, or πD.
The area is π x Radius x Radius, where the Radius is half of the Diameter.
The area formula is also written πR^2.

a. If a circle has circumference equal to 12π cm, what is the area of the circle in terms of π?

Solution: If the circumference is 12π cm, then the diameter of the circle is 12 cm. That means the radius of the circle is 6 cm. Therefore the area of the circle is 36π square cm.

b. If a square has perimeter equal to 12π, what is the area of the square?

Solution: If the square has perimeter 12π, then the length of one side is 3π. (12π divided by four.) The area of the square is 3π x 3π = $9\pi^2$ square cm.

c. Which has the larger area, the circle with perimeter 12π, or the square with perimeter 12π?

Solution: The area of the circle with circumference 12π is 36π, which is about 113 square cm. The area of the square with perimeter 12π is $9\pi^2$ which is about 89 square cm. So the circle encloses significantly more area than the square of corresponding perimeter.

7-8. Bicycle Wheel

If an 80 cm diameter bicycle wheel turns 2 full revolutions per second, how fast is the bicycle moving in kilometers per hour?

Information you need:
a. The circumference of the wheel is π x diameter.
b. If the wheel turns 2 full revolutions per second, how many circumferences does it travel per second?
c. The speed is the distance traveled ÷ time.
d. There are 100 centimeters in 1 meter.
e. There are 1000 meters in 1 kilometer.
f. There are 3600 seconds in 1 hour.

Solution: The circumference of the wheel is π x 80 cm.
The distance that the wheel travels in one second is two revolutions:

2 x π x 80. Therefore the speed is 2 x π x 80 = 160π cm/second.

To convert that speed into kilometers per hour:

$$160\pi \frac{cm}{\sec} \times \frac{1\ meter}{100cm} \times \frac{1kilometer}{1000meters} \times \frac{3600\sec onds}{1hour} = 5.76\pi \frac{km}{h} \approx 18.1 \frac{km}{h}$$

The bicycle moves approximately 18.1 kilometers per hour.

7-9. Triangle Areas

Inside every rectangle are two triangles. They are called right triangles because one of the angles in the triangle is a right angle. The area of a right triangle is easy to calculate. It's half of the area of the rectangle that the right triangle sits in.

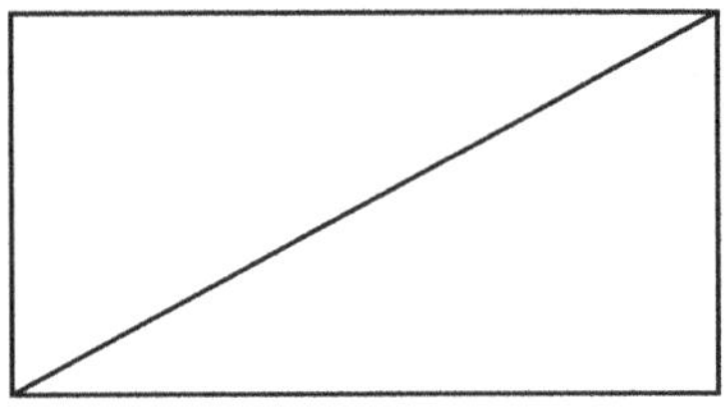

This idea extends to the area of any triangle. If a random triangle is not a right triangle, then you can always rotate it so that one of the sides is horizontal, and the highest point is over the base.

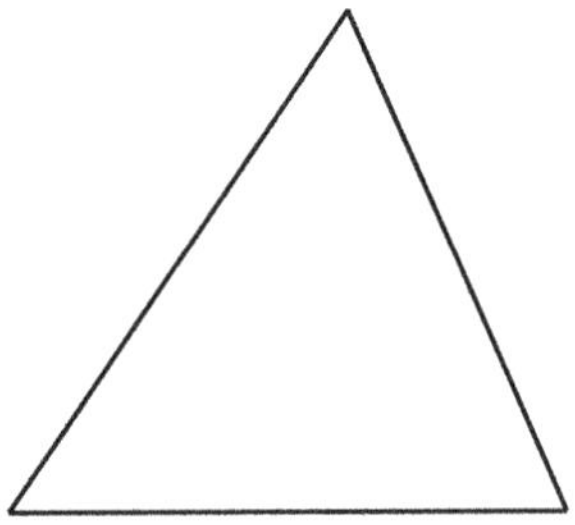

Show that the area of the triangle below is ½ x Base x Height. Use the two rectangles shown.

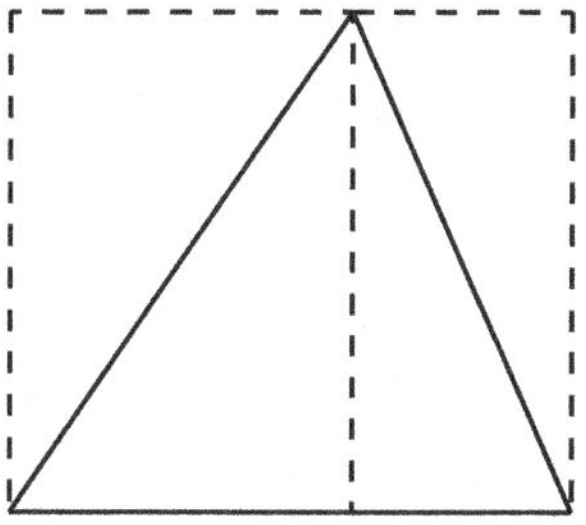

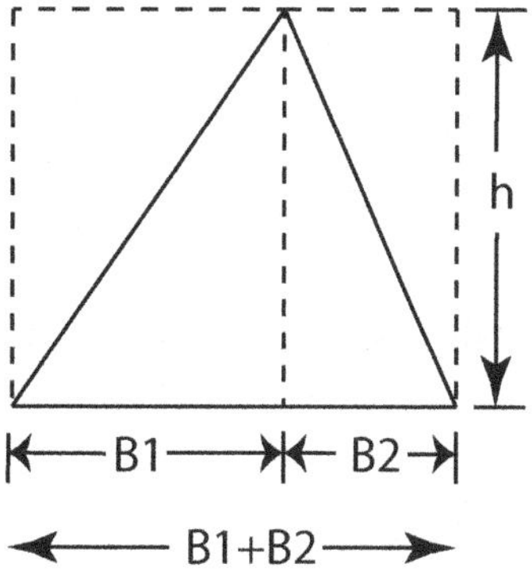

Solution: Label the bases and height of each triangle in the diagram, as above: The area of the full triangle is the sum of the areas of the smaller right triangles: ½ B1 x h + ½ B2 x h

This is the same as ½ (B1 + B2) x h
If the students have a hard time with this, they can do experiments with real numbers.

Thus the area of the triangle is ½ Base x Height.

7-10. Cover Squares

On the cover of this book are four diagrams of squares within a square. In the first diagram, there are 5 squares, in the second, 25, and so on.

a. What is the total area of the squares in each diagram? (Four answers)

Note to Coach: This is a good moment to discuss units of measure with your students. The largest black squares on the cover measure about 27 millimeters on a side, a little more than an inch. The nice thing about 27 is that it divides evenly into three, three times, so the next size squares are 9 mm, 3 mm, and 1 mm.

Alternatively, you could tell your students to simply assume that the largest squares are 1 unit long on a side, and then smaller squares are 1/3 unit, $1/9^{th}$ unit, and $1/27^{th}$ of a unit.

Depending on whether you'd like to use real measurements or relative units, you can go either way.

However you decide to treat the units of measurement, the point should not be lost that the area diminishes, while the perimeter skyrockets.

Solution: First diagram: side length 1 unit. Number of squares: 5. Total area is 5 square units. In cm, side length is 2.7 cm, total area is 36.45 cm^2.

Second diagram: side length 1/3 unit. Number of squares: 25.
Area of one square: 1/9 square units. Total area: $25/9^{th}$ square units.
In cm, the side length is 0.9 cm, area of one square is 0.81 cm^2.
Total area: 20.25 cm^2.

Third diagram: side length is 1/9 unit. Number of squares: 125.
Area of one square: 1/81 square unit. Total area: 125/81 square units.
In cm, side length is 0.3 cm. Area of one square: 0.09 cm^2.
Total area: 11.25 cm^2.

Fourth diagram: side length is 1/27 unit. Number of squares: 625.
Area of one square: 1/729 unit. Total area: 625/729 square units.
In cm, side length is 0.1 cm. Area of one square: 0.01 cm^2.
Total area: 6.25 cm^2.

b. What is the total perimeter of the squares in each diagram? (Four answers)

Solution: First diagram: side length is 1 unit. Number of squares: 5.
Perimeter of one square: 4 units. The total perimeter is 20 units.
In cm, side length is 2.7 cm. The total perimeter is 54 cm.

Second diagram: side length is 1/3 unit. Number of squares: 25.
The perimeter of one square: 4/3 unit. The total is 100/3 units.
In cm, side length is 0.9 cm, perimeter of one square: 3.6 cm, and total perimeter is 90 cm.

Third diagram: side length is 1/9 unit. Number of squares: 125.
Perimeter of one square: 4/9 unit. Total perimeter: 500/9 units.
In cm, side length is 0.3 cm, perimeter of one square: 1.2 cm, and total perimeter: 150 cm.

Fourth diagram: side length is 1/27 unit. Number of squares: 625.
Perimeter of one square: 4/27 unit. Total perimeter: 2500/27 units (about 92.6 units).
In cm, side length is 0.1 cm, perimeter of one square: 0.4 cm, and total perimeter: 250 cm. (Two and a half meters. Taller than a person.)

Note to Coach: It's a fun idea to invite comparisons to these lengths. What are some things that are 150 cm long? 90 cm? 54 cm?

How To Deal With Area and Perimeter Problems

The following are only suggestions:

Learn the vocabulary:

- Dimension
- Area
- Length
- Width
- Perimeter
- Units
- Shape
- Rectangle
- Triangle
- Right triangle
- Circle
- Circumference
- Pi
- Diameter
- Radius

Look for patterns such as:

- Problems asking for multiple rectangles of a given area are asking for factors of that area.
- Every rectangle is made up of two right triangles joined at the diagonal.

There might be surprises such as:

- Two very different shapes can have the same area.
- A very small area can be contained in a very long rectangle.
- Areas are much harder to estimate than lengths.

8. Probability

Why Probability

Here are Professor Bear's reasons to for the eighth chapter to be about Probability:

- Probability is an example of a mathematical model: the numbers from 0 to 1 are assigned a meaning. Sets and subsets, unions and intersections are assigned a meaning. Many probability problems involve interpreting numbers in the context of a model.
- Probability problems are applications of arithmetic and fractions.
- Probability uses set theory.
- Probability uses measure theory: how to measure the size of a set. Because of the probability model, a lot of mathematics has been invented to explore probability. In particular, a lot of mathematics having to do with measuring the size of a set.

Underpinnings of Probability

To understand this material students need some experience with arithmetic and should be able to do one and two digit addition, subtraction, multiplication and division.

They need to know fractions: simplest terms, equivalent fractions, adding subtracting multiplying and dividing fractions.

It's helpful to know sets and subsets.

Vocabulary is very important in probability. Chance, outcome, event, likely, unlikely, impossible, sure, certain—these are all words that can be used in probability problems.

Outcomes

After doing this chapter students will have experience with the idea of probability and learn how to calculate some simple discrete probabilities.

Probability: Notes and Vocabulary

An **event** is something that could happen. The **probability** of an event is a number in the range from 0 to 1 that expresses the likelihood of the event happening.

If the probability is 1, the event will definitely happen.
If the probability is 0, the event would never happen.
If the probability is somewhere between 0 and 1, the event might happen. The closer the probability is to 1, the more likely the event will happen.

The probability is this fraction: $\dfrac{\textit{number of ways event could happen}}{\textit{total number of events}}$

If you flip a coin once, there are 2 possible events: it comes up Heads or it comes up Tails.

Coming up Heads is 1 event. The probability that it comes up Heads is $\frac{1}{2}$.

The set of all possible events is called the **sample space**. In the case of flipping a coin once, the sample space is the set {comes up Heads, comes up Tails} which we can abbreviate as {H, T}.

One important part of probability is therefore counting. The number in the denominator of the probability is the number of objects in the sample space, so you need to count them.

Example: If you flip two coins, what is the sample space?

Answer: {HH, HT, TH, TT}. The sample space is the set of all possible results of flipping two coins. The size of this sample space is 4.

Another example: if you toss one 6-sided die, what is the sample space?

Answer: {1, 2, 3, 4, 5, 6} The size of this sample space is 6.

If you flip two coins, here some possible events and their probabilities:

Knowing the sample space makes it easier to calculate probabilities.

In the case of flipping two coins, the following table shows some events and how to calculate their probabilities.

Event	**Probability of Event, given sample space of {HH, TH, HT, TT}**
Two Heads	The number of ways HH occurs in the sample space is 1. Therefore: P(HH) = ¼
One Head, one Tail	There are two ways to achieve one Head and one Tail: HT, TH. P(HT or TH) = 2/4 = ½
Two Tails	There is one way to achieve two Tails: TT P(TT) = ¼
At least one Tail	There are three ways to achieve at least one Tail: HT, TH, TT P(HT or TH or TT) = ¾
Both coins disappear and go to Jupiter	This is so unlikely to happen that it is not in our list of possible events in the sample space. P(disappearing and going to Jupiter) = 0/4 = 0

8-1. Sample Space

Each of these probability questions has a related sample space. What is the sample space? What is the size of the sample space?

Note to Coach: This may seem like a strange problem. Usually we're just interested in probabilities, so why does the problem ask for sample spaces and sizes of sample spaces. The reason is: it's usually impossible to get the correct probability without knowing the size of the sample space. And, many probability problems do not spell out the sample space; you have to infer the sample space from the statement of the problem. So these problems give students some practice to do that. The problems are worded as probability problems, but students must read carefully to figure out the full sample space.

a. Mrs. Cumberbatch is going to buy a new car. She'll choose the color at random. The available colors are: red, silver, white, and black. What is the probability she'll get a silver or black car?

Solution: The sample space is the full set of colors: red, silver, white, black. The size of the sample space is 4.

The probability she'll get a silver or black car, two possibilities out of four, is ½.

b. What is the probability that the person standing closest to you was born on Thursday?

Solution: The sample space is the set of days of the week: Monday, Tuesday, Wednesday, Thursday, Friday, Saturday, Sunday. The size of the sample space is 7.

The probability that the person closest to you was born on Thursday is one possibility out of seven, 1/7.

c. What is the probability of tossing three Heads in a row (with a fair coin)?

Solution: The sample space is the set of all three tosses. Let H stand for Head and T stand for Tail. Then the sample space is:
HHH, HHT, HTH, THH, HTT, THT, HHT, TTT
The size of the sample space is 8.

The probability of tossing three Heads in a row, HHH, is one out of eight, 1/8.

d. A computer generates numbers at random from the four digits 4, 5, 6, and 7. What is the probability that it generates the number 7654? (Each digit is used once.)

Solution: Since each digit is only used once, our sample space does not contain numbers such as 4444. The sample space is the set of all **permutations** of the digits 4, 5, 6, and 7. (The set of all ways of mixing up the four digits.)

There are two ways to mix up two numbers A and B: AB and BA. If there is a third number, C, then there are a total of six possibilities:
Begin with A, and mix up B and C (ABC, ACB)
Begin with B, and mix up A and C (BAC, BCA)
Begin with C, and mix up A and B (CAB, CBA)

If you throw in a fourth number, D, then there are four times six possibilities:
Begin with A, and mix up B, C, and D
Begin with B, and mix up A, C, and D
Begin with C, and mix up A, B, and D
Begin with D, and mix up A, B, and C

Therefore, there are twenty-four permutations of 4, 5, 6 and 7.

e. Mrs. Cumberbatch went back to the car store. She found out she can get one of two kinds of tires: radial, or all-weather, and one of three kinds of window tinting: gray, aqua, or taupe. What is the probability she gets all-weather tires and taupe tinting?

Solution: The sample space for this problem consists of pairs of tires and tinting:
radial + gray, radial + aqua, radial + taupe,
all-weather + gray, all-weather + aqua, all-weather + taupe
The size of the sample space is 6, and the probability she gets all-weather tires with taupe tinting is 1/6.

8-2. Gumballs

Mr. Grumby wants to buy four gumballs of the same color from a gumball machine. The machine contains 6 yellow, 7 red, and 8 blue gumballs. How many gumballs does Mr. Grumby have to buy to be certain he has four gumballs of the same color?

Solution: One approach to this problem could be called "worst-case scenario." If Mr. Grumby were most unlucky, he'd keep buying gumballs of different colors, for example: yellow, red, blue, yellow, red, blue, yellow, red, blue. However, once he'd bought three of each color (a total of nine), he'd be forced to buy a fourth gumball of one color. So he'd be certain to have at least four gumballs of the same color if he bought ten gumballs. There are a total of 21 gumballs in the machine so buying ten of them is possible.

Another approach might be "color buckets". Suppose every time Mr. Grumby buys a gumball, he puts it in a matching color bucket. Once Mr. Grumby buys four gumballs of the same color, he stops. If Mr. Grumby keeps going and buys nine gumballs, each color bucket would have to have three gumballs of each color (this needs to be thought through). But once Mr. Grumby bought the tenth gumball, one of the color buckets would have to have four balls in it.

Observation: There is an interesting behind-the-scenes problem within this problem. Another way to state the problem is this: how many balls would Mr. Grumby have to buy such that every element in the sample space of ball-combinations always contains at least four gumballs of the same color?

Clearly, if Mr. Grumby buys fewer than four gumballs, there is no chance of four gumballs of the same color. What does the sample space look like for four gumballs? For six gumballs? Nine gumballs? Ten gumballs? The sample spaces are quite large. Exploring the sample spaces gives a lot of practice in counting.

Here is the sample space for four gumballs: YYYY, RRRR, BBBB, YYYR, YYYB, RRRY, RRRB, BBBY, BBBR, YYRR, YYBB, RRBB.

8-3. Sack Of Balls

A sack contains balls that are identical except for color. There are 6 red balls, and the remaining balls are all green. If the probability of drawing a red ball at random from this sack is ¼, how many green balls are in the sack?

Solution: The probability of drawing a red ball is 6/(total number of balls). This becomes a ratio problem: 6:total = 1:4. There must be a total of 24 balls in the sack. If 6 are red, 18 balls are green.

8-4. Six-Sided Die

A six-sided die is thrown twice.

a. What is the probability that the sum of the throws is 4?

Solution: This problem is equivalent to throwing two dice once. The sample space of outcomes has thirty-six elements, depending on the outcomes of the throws: one-one, one-two, ... one-six, two-one, and so on (see the table below). The only combinations that sum to 4 are: one-three, three-one, and two-two. These are 3 out of 36 possibilities. Therefore the probability is 1/12th.

b. What is the probability that the sum of the throws is divisible by 3?

Solution: An explicit way to determine the outcomes divisible by 3 is to make a grid of all the outcomes:

1-1	1-2	1-3	1-4	1-5	1-6
2-1	2-2	2-3	2-4	2-5	2-6
3-1	3-2	3-3	3-4	3-5	3-6
4-1	4-2	4-3	4-4	4-5	4-6
5-1	5-2	5-3	5-4	5-5	5-6
6-1	6-2	6-3	6-4	6-5	6-6

If you circle the outcomes divisible by 3, they form a pattern on the grid and are easy to count. They are 12/36. Therefore the probability is 1/3rd.

8-5. Tantrum Coin

Bob and George are coins. If you flip Bob any time, the probability that he turns up Heads is ½. If you flip George alone, the probability that he turns up Heads is ½. But if you flip Bob and George together, George has a tantrum and only turns up Tails.

a. Write down the set of possible events if Bob is tossed twice, and then George is tossed twice after Bob.

Solution: HHHH, HTHH, THHH, TTHH,
HHTH, HTTH, THTH, TTTH
HHHT, HTHT, THHT, TTHT
HHTT, HTTT, THTT, TTTT

The sample space has 16 elements.

b. Write down the set of possible events if Bob and George are tossed together, twice.

Solution: HHTT, HTTT, THTT, TTTT

The sample space has 4 elements.

c. Bob and George's parents prefer Heads to Tails. They award Bob and George one team point for every Head that appears.

i. What is the probability that Bob and George earn 2 points, if they are each flipped twice, separately?

Solution: Looking at the sample space of part a, there are six possibilities out of sixteen where there are two heads. So the probability is 3/8th.

ii. What is the probability that Bob and George earn 2 points, if they are flipped twice, together?

Solution: Looking at the sample space of part b, there is one possibility of two heads, out of four total. So the probability is ¼ .

iii. What is the probability that Bob and George earn no points, if they are each flipped twice, separately?

Solution: Looking at the sample space of part a, there is one possibility of TTTT out of sixteen. So the probability is 1/16th.

iv. What is the probability that Bob and George earn no points, if they are flipped twice, together?

Solution: Looking at the sample space of part b, there is one possibility of TTTT out of four. So the probability is ¼ .

Extension: It is interesting to bring up the concept of expected value. You can ask the question: how many points, on average, do you expect George and Bob to earn if they are tossed twice?

In case a, where George does not throw a tantrum, the expected value is: $4\times\frac{1}{16}+3\times\frac{4}{16}+2\times\frac{6}{16}+1\times\frac{4}{16}=2$ (The sum of the point values weighted by their probabilities; for example the probability of getting 4 points is 1/16.)

In case b, where does throws a tantrum, the expected value is: $2\times\frac{1}{4}+1\times\frac{1}{2}=\frac{3}{4}$

8-6. Card Draw

Royce has three cards: an Ace, a King, and a Queen. What is the probability that if Royce draws two cards at random, one of the cards is the Ace?

Solution: Let the Ace = A, King = K, Queen =Q. The sample space for this problem is AK, KA, AQ, QA, KQ, QK. The probability that one card is an Ace is 4/6 = 2/3.

8-7. Random Number Generator

If a computer generates random numbers using the four digits 4, 5, 6, and 7, what is the probability that it generates a number less than 5467? (Refer to Problem 1 d).

Note: each digit is used once.

Solution: The key to solving this problem is to observe that 5467 is the lowest number in the sample space that begins with 5. That means that the numbers we are looking for all begin with 4 (4 is the thousands digit). In other words, the problem is asking for the probability that the random number begins with 4.

If students have a hard time seeing this, it's worthwhile to write down some elements of the sample space.
The smallest possible number is 4567, followed by 4576. What's the next largest number? What's the largest possible number beginning with 4? What's the next largest number after that? (5467).

Each random number in this problem is a permutation of the digits 4, 5, 6, and 7. In previous Counting problems (such as 2-8b) the number of permutations of four objects is 24. So the sample space has size 24. The number of numbers that begin with 4 is six (4567, 4576, 4657, 4675, 4756, 4765). So the probability is 6/24 = ¼ .

Once you realize the problem is asking what is the probability that the random number begins with 4, you can quickly come to the solution of ¼, because there are four possible numbers to begin with (4, 5, 6, 7) and 4 is one of them.

8-8. Independence

Independent events are not affected by other events. Dependent events might be affected by other events. If events have some sort of dependence, the probabilities would change. An example of dependence is the coin in Problem 5 that throws a tantrum. In the following questions, think about whether the events described are independent or dependent.

a. Larry tosses a fair six-sided die four times. The results of the tosses are 2, 6, 4, and 2. What is the probability that the die will come up 5 on Larry's fifth toss?

Solution: The probability is 1/6. If the die is fair, all sides have equal probability for every toss.

b. Quarole tosses a fair six-sided die four times. All four times, the die comes up 5. What is the probability that the die will come up 5 on Quarole's fifth toss?

Solution: The probability is 1/6. If the die is fair, all sides have equal probability for every toss.

c. Spellman has a plastic die that comes up 5 with a probability of 1/6. But then Spellman leaves the die in his toaster and it melts, so that the 5-face is now together with the 6-face, making an 11-face. The other faces are the same as before. What is the probability that the die comes up 5 after being in the toaster?

Solution: In this case, the die has changed. So the previous probabilities no longer apply. The answer to the problem is zero, because after melting, the 5 cannot show up by itself anymore.

d. Harry does not want new shoes. Dudley gets new shoes and brags about how dry and warm his feet feel. What is the probability that Harry now wants new shoes?

Solution: There is no specific solution to this problem, in our opinion. Students can answer the problem according to their own experience, but they should be prepared to explain their conclusions. Professor Bear's answer, based on her experience, is the following: Harry might or might not want new shoes after seeing and smelling Dudley's. But some people want new shoes whenever they see them. So I believe there is a small, positive (nonzero) probability that Harry now wants new shoes. The exact value of that probability is best estimated by experiments.

Another person might state, from his experience: the probability is zero. One person having new shoes never forces another person to want new shoes.

e. Harry is allergic to peanuts. Harry sees Dudley eat a whole peanut butter cream pie with great enjoyment. What is the probability that Harry is still allergic to peanuts?

Solution: This problem is different from the previous one. The probability is zero. Seeing Dudley eat a peanut product has no effect on Harry's allergies.

8-9. Favorite Foods

Lori conducted a survey of all fourth-graders at her school. She asked each of them to name their favorite food. Here is her data:

Food	Number of Kids
chow mein	8
cookies	6
cupcakes	3
French fries	4
hamburgers	2
ice cream	6
pizza	13
spaghetti	8
no favorite	10

a. What is the size of Lori's sample space?

Solution: The sample space is the set of responses from kids. The size is 8 + 6 + 3 + 4 + 2 + 6 + 13 + 8 + 10 = 60

b. What is the probability that a random fourth-grader from Lori's school has spaghetti for his or her favorite food?

Solution: In Lori's sample, 8/60, or 2/15 students prefer spaghetti. So the probability that a random fourth-grader from Lori's sample favors spaghetti over other foods is 2/15. Lori's data is only about the 60 students that she surveyed. So if it were a year later, the probability might not be valid.

c. What is the probability that a random fourth-grader from Lori's school has a dessert item for his or her favorite food?

Solution: The dessert items are cookies, cupcakes, and ice cream. The number of students who like desserts the most is 6 + 3 + 6 = 15. Therefore the probability that a random fourth-grader from Lori's sample favors a dessert is 15/60 or ¼. Again, the data is only valid for Lori's sample.

d. Lori wants to find out the probability that the favorite food of a random fourth-grader in the United States is ice cream. What should she do?

Solution: This is an open-ended question. It is time for students to think about the problem and come up with strategies. Some examples are:

- Get more data somehow, and estimate the probability.
- Calculate the probability from her own sample (1/10) and go to the library and see if there are other surveys of fourth graders that ask about ice cream. See if they have comparable results.
- Learn more about probability theory. If they dig a little deeper they will find out that probability theory can provided answers to questions such as "how many kids should Lori sample in order to get an accurate value of probability?"

8-10. Probability And Statistics

How are statistics estimated using probability? This subject is way beyond this book, but it is so widely used that it is worth mentioning. In many math classes Statistics and Probability are studied at the same time. You can see from Problem 9 how Statistics are used to estimate probabilities and it turns out that probability is used to estimate statistics.

It is almost always impossible to get *all* the data you need for accurate statistics. Suppose, for example, you want to know the average weight of a rabbit in the United States. To answer that question correctly, you'd need to gather the weights of every rabbit in the United States. Getting all that data would be difficult.

Probability can help. Probability can answer the question: how many rabbit-weights do I need in order for me to get a "pretty good" answer.

If you gather the weights of all the rabbits, the probability that your average is correct is 1. If you gather the weights of zero rabbits, the probability that your average weight is correct is zero. Somewhere in between all the rabbits, and zero rabbits is enough rabbits so that the *probability that your average weight is correct* is 0.95, or 95%.

In statistics, we call this a 95% confidence level.

What statisticians do is estimate the number of rabbits needed for 95% using a mathematical model of the rabbit data.

Most of the time, when people are talking about statistics or statistical analysis, they are talking about using statistics and probability together in exactly this way.

If the statistic in question is extremely sensitive or important, we might want to get enough data to reach a 99.9% confidence level. If we just need a general ballpark estimate, we might settle for a 70% confidence level.

Here is a question to think about:

A and B claim to know the average allowance of students at Bellinger High School. A says: "The average allowance is $7.68 per week. I sampled 89 students and my confidence level is 98%." B says: "The average allowance is $6.92 per week. I sampled 145 students." Who has the more accurate average, A or B? Put another way, which number do you believe more, A's number or B's number?

Solution: This is another open ended problem. The solution is somewhat open to debate.

If A and B both take equal care to find a random sample of students, then B's result should be closer to reality, because B sampled more students. The most accurate result would come from a sample space of the whole school. The closer the sample space size to the size of the whole school, the better the result (why? If students don't understand this, they can experiment.)

On the other hand, the fact that A provides a confidence level of 98% and B provides no confidence level at all, makes Professor Bear lose a little confidence in B. Professor Bear would want to know if B had done as good a job as A in gathering the data. What if B had surveyed more students, but B had only surveyed her friends? What if B had only surveyed students who complained about their allowance?

Implicit in A's guarantee of 98% confidence level is some assurance that A tried to get good data. B might have tried just as hard. From the statement of the problem, we don't know.

It might be interesting to look for examples in advertisements, product descriptions, or news, that quote confidence levels, and ask students what they think the confidence level means.

How To Deal With Probability Problems

The following are only suggestions:

Learn the vocabulary:

- Sample space
- Probability
- Event

- Independent events
- Dependent events

Look for patterns such as:

- In determining the size of a sample space, multiplication of possibilities can be involved.
- If double possibilities are allowed, for example HH or rolling snake eyes with dice, then straight multiplication of the possibilities works.
- If you are counting permutations of a fixed set of objects, and objects are not repeated, then the number of permutations is N! (N factorial).

There might be surprises such as:

- The probability of tossing a head after tossing ten heads is still ½.
- The probability that the sum of a roll of two dice is divisible by 3 is 1/3rd.

9. Bases

Why Bases

Here are Professor Bear's reasons to for the ninth chapter to be about Bases:

- This chapter is about number bases other than 10: binary, ternary, hexadecimal to name a few.
- Adding, subtracting, multiplying and dividing all have the same algorithms in different number bases. Applying these algorithms in a different base is a wonderful way to deepen one's understanding of what these actions are.
- A big topic in elementary school is place value. Actually it's a big stumbling block for many. Seeing how it works in bases other than 10 can help form a deeper understanding of place value and remainders.
- Binary, octal, hexadecimal, and base-64 numbers are used extensively in computer programming.
- This chapter appears toward the end of the book because it is seldom covered in math contests, and requires considerable comfort with regular number operations.

Note to Coach: this chapter defines binary numbers with a few initial addition rules, such as 1 + 1 = 10.

When adults learn about binary numbers, place value might be the best place to start. A binary number has a ones place, twos place, fours place, and so on. Place value requires a good understanding of powers, exponents, and reciprocals.

If, however, the student has not learned place value yet, it can be very confusing. It can be better to start as a game, where the rule of the game is that 1 + 0 = 1, and 1 + 1 = 10 (carry the one, or use one's favorite addition algorithm). That is the approach used in the Workbook. Binary numbers can be "built" from smaller binary numbers by simple adding and multiplying.

It's interesting to compare "long addition" in binary vs. "long addition" in decimal. In decimal, students have to learn a series of addition facts: single digit sums. In binary, the situation is much easier. There are three facts: 0 + 0 = 0, 1 + 0 = 1, and 1 + 1 = 10.

The tradeoff is that binary numbers are much longer and require more "carrying". It's worth asking students what some other tradeoffs might be—when might binary numbers be an advantage over decimal numbers?

Underpinnings of Bases

To understand this material you need some experience with arithmetic and be able to do one and two digit addition, subtraction, multiplication and division. You need to be able to read a graph.

Remainders are very important. All place-value style numbers are strings of remainders: for example in base 10, the number 54321 has a remainder of 1 when divided by 10; a remainder of 21 when divided by 100; a remainder of 321 when divided by 1000, and so on.

You need fractions and place value.
You also need divisibility rules.

Outcomes

After doing this chapter you'll be able to convert decimal numbers to another base, and work out basic arithmetic in the other base. You'll see some of the ways binary numbers are useful in computing, and have a way to make your own secret code numbers by converting to a different base.

Bases: Notes and Vocabulary

This chapter is about how to "spell" numbers in different ways. It turns out that the way we "spell" numbers can show off different kinds of patterns and behavior.

You already know, for example, that numbers such as 20, 450, 670, 1000, and 390 are all divisible by 10. That is because in base ten notation, these numbers all end with 0.

The **notation** of a number is how we "spell" it. Here are two notations of the same number: five and 5.

There are many other ways to notate the number five.

Base ten notation uses the ten Arabic numerals 0, 1, 2, 3, 4, 5, 6, 7, and 9. You already know a lot about base ten notation. You know that the right-most digit is the ones digit, and the digit to the left of the ones digit is the tens digit, and so on.

Base ten notation is also called **decimal notation**, or decimal for short.

Base two notation uses the two Arabic numerals 0 and 1. Base two notation is also called binary notation or binary for short.

You can convert any decimal number to a binary number, and vice versa.

First, to convert whole numbers:

Zero is the same in decimal and binary: 0.
One is the same in decimal and binary: 1.

1 + 1 = two in both decimal and binary. However in binary there is no digit 2. Once you get to two in binary, you run out of digits. When do you run out of digits in decimal?

At number nine. The next whole number, ten, is notated 10 in decimal.

In binary, two is notated 10. One way to think about the meaning of "10" as two is place value. In any base, the right-most digit is the ones digit. In base ten, the next digit is the tens digit. In base two, the next digit is the *twos digit*. The number two is 1 two, 0 ones: 10.

You can write equations in binary:

$1 + 1 = 10$ or $\begin{array}{r} 1 \\ +1 \\ \hline 10 \end{array}$ see how you "carry 1's" just like in base ten.

By using the rules you already know about addition, you can write the first several binary whole numbers. Fill in the table below. (Solutions are added).

Number	Binary Sum	Binary Number
two	1 +1 — 10	10
three	10 +1 —	**11**
four	10 11 +10 or +1 —— — 100	100 The next digit in base two is the fours digit!
five	11 100 +10 or +1 —— ——	**101**
six	11 100 +11 or +10 —— ——	**110**
seven	100 +11 or ——	**111**
eight	100 +100 or ——	**1000**
nine	110 +11 or ——	1001

In the rest of this chapter we'll explore binary arithmetic and other bases. It's important to remember that much of the math you already know—such as long multiplication and long division—applies to numbers in base 2 and other bases. Here are some binary math questions to get started. Assume all of the numbers on this page are in base 2.

10 – 1 = **1**

100 – 10 = **10**

10 x 10 = **100**

Describe the difference between 1110 and 111.

Solution: There are a couple of approaches. One is to immediately notice that 1110 is 10 (two) multiplied by 111 (seven). Another approach is to convert each number to decimal: 1110 is fourteen, and 111 is seven, so the longer number is double the shorter.

Describe the difference between 11100 and 111.

Solution: 11100 is 16+8+4 = 28 in decimal. So 11100 is four times 111.

How do you write the double of a binary number? For example, how do you write the double of 11010101?

Solution: Add a zero to the right end. The double of 11010101 is 110101010.

How can you tell if a binary number is odd?
Solution: Its right-most digit is 1.

How can you tell if a binary number is even?
Solution: Its right-most digit is 0.

Fill in the table: (Solutions entered).

Number	Tens Digit in decimal	Ones Digit in decimal	Eights Digit in binary	Fours Digit in binary	Twos Digit in binary	Ones Digit in binary
nine	**N/A**	**9**	**1**	**0**	**0**	**1**
twelve	**1**	**2**	**1**	**1**	**0**	**0**
fifteen	**1**	**5**	**1**	**1**	**1**	**1**

Solutions: Problems

9-1. Binary Adding And Subtracting

Try some binary addition and subtraction. Convert the problems to decimal and check your answers.

Solution:

	101		1111		1111		1000
a.	+10	b.	+1010	c.	−1010	d.	−101
	111		11 001		101		11

a. 5 + 2 =7 b. 15 + 10 = 25 c. 15 − 10 = 5 d. 8 − 5 = 3

9-2. Binary Long Multiplication

Try some binary long multiplication. Convert the problems to decimal and check your answers. The first problem is done as an example.

Solution:

	111				110		1001
	×11		1010		×101		×11
a.	111	b.	×10	c.	110	d.	1001
	1110		10100		11000		10010
	10101				11110		11011

a. 7x3 = 21 b. 10 x 2 = 20 c. 6 x 5 = 30 c. 9 x 3 = 27

9-3. Binary Sequences

Sometimes it is easier to spot a sequence pattern in binary than in decimal, or vice versa. In each of the conversion problems below, ask yourself, what is the pattern of the sequence? Is it easier to see in decimal or in binary?

a. Convert the sequence 1, 2, 4, 8, 16, . . . to binary.

Solution: 1, 10, 100, 1000, 10000

b. Convert the sequence 5, 9, 17, 33, . . . to binary.

Solution: Here is an algorithm for converting to binary: Divide successively by two, and note down the remainders.

5 divided by two has quotient 2, remainder 1. The ones digit of 5 is 1.

Quotient 2 divided by two is 1, remainder zero. So the twos digit of 5 is 0 and the fours digit of 5 is 1. In binary, 5 is 101.

9 divided by two is 4, remainder 1. The ones digit is 1.
4 divided by two is 2, remainder 0. The twos digit is 0.
2 divided by two is 1, remainder 0. The fours digit is 0, and the eights digit is 1. In binary, 9 is 1001.

17 divided by two is 8 remainder 1. The ones digit is 1.
8 divided by two is 4, remainder 0. The twos digit is 0.
4 divided by two is 2, remainder 0. The fours digit is 0.
2 divided by two is 1, remainder 0. The eights digit is 0, and the sixteens digit is 1. In binary, 17 is 10001.

Through a similar process, you can see that in binary, 33 is 100001.

A quicker way to convert the numbers to binary is to observe that each number is one more than a power of two.

5 = four + 1	In binary, 5 = 100 + 1 = 101
9 = eight + 1	In binary, 9 = 1000 + 1 = 1001
17 = sixteen + 1	In binary, 17 = 10000 + 1 = 10001
33 = thirty-two + 1	In binary, 33 = 100000 + 1 = 100001

Thus the sequence is 101, 1001, 10001, 100001, ...

c. Convert the binary sequence 1, 100, 111, 1010, 1101, 10000, . . . to decimal. What is the pattern?

Solution: Converting to decimal:
1 is 1
100 is 4
111 is 100 + 10 + 1 which is four + two + one = seven = 7
1010 is 1000 + 10 which is eight + two = ten = 10
1101 is 1000 + 100 + 1 = eight + four + one = thirteen = 13
10000 is sixteen = 16
The sequence is 1, 4, 7, 10, 13, 16, . . .
And the pattern is to increase by three.

9-4. Divisibility In Binary

In the Factors and Primes chapter we discussed divisibility rules for decimal numbers. Do the same rules apply in binary?

a. How can you tell if a binary number is divisible by two?

Solution: In binary, two is 10. Any product of two ends in 0 in binary. Any number divisible by two must end in 0 (zero).

b. The divisibility rule for three in decimal is: the number is divisible by three if the digits add up to a number divisible by three. Does the same rule apply in binary?

Solution: No. The simplest example is three. In binary, three is 11. The sum of the digits is two, which is not divisible by three. So the "sum of the digits rule" for divisibility by three does not work.

c. How can you tell if a binary number is divisible by four?

Solution: In binary, four is 100. Any product of four ends in 00 in binary. Any number divisible by four must end in 00. (Double check with long multiplication).

9-5. Place Value In Binary

Binary numbers, like decimal numbers, have place values. A three-digit decimal number such as 457 has a ones digit, a tens digit, and a hundreds digit. The maximum three-digit decimal number is 999, nine hundred and ninety-nine.

a. What are the three place values in a three-digit binary number?

Solution: The three place values are powers of two. They are: ones place, twos place, fours place.

b. What is the maximum three-digit binary number? Convert this number to decimal.

Solution: The maximum three-digit binary number is 111.
111 = 100 + 10 + 1 = 4 + 2 + 1 = 7

c. What is the maximum four-digit binary number? Convert this number to decimal.

Solution: The maximum four-digit binary number is 1111. Using the answer from part, b:
1111 = 1000 + 111 = 8 + 7 = 15

d. What are the five place values of a five-digit binary number?

Solution: Ones place, twos place, fours place, eights place, sixteens place.

e. What is the maximum five-digit binary number?

Solution: The maximum five-digit binary number is 11111. In decimal:
11111 = 10000 + 1111 = 16 + 15 = 31

Alternate: Notice that the number 11111 in binary plays a similar role to 99999 in decimal. 99999 is one less than 100000. In binary, 11111 is one less than 100000. Therefore:
11111 = 100000 – 1 = 32 – 1 = 31

This same alternate solution works for parts b and c.

f. (Read). Convert the decimal number 11 to binary using the following method:
What is the highest power of 2 less than 11 (is it 2, 4, 8, 16, 32)? 8
Subtract 8 from 11 and you get a remainder of: 3
The highest power of 2 less than 3 is 2.
Subtract 2 from 3 and you get a remainder of 1.
Therefore, 11 = 8 + 2 + 1.
In binary, eleven = 1000 + 10 + 1 = 1011.

Alternate: Use the algorithm described in the solution of 9-3: Divide successively by twos and note remainders; the final 1 is the last (highest) digit.
11 divided by two is 5 remainder 1. The ones digit is 1.
5 divided by two is 2 remainder 1. The twos digit is 1.
2 divided by two is 1, remainder 0. The fours digit is 0.
We reached the final 1. The eights digit is 1.
Therefore eleven in binary is 1011.

Observe: A number expressed in any base is a succession of remainders. Consider the number 12345 in decimal.
The remainder when you divide 12345 by 10 is 5.
The remainder when you divide 12345 by 100 is 45.
The remainder when you divide 12345 by 1000 is 345.
And so on.

This algorithm works when converting to any base.

g. Convert the decimal number 26 to binary using the same method as above:

What is the highest power of 2 less than 26? **16**
Subtract this power of 2 from 26 and you get a remainder of: **10**
The highest power of 2 less than the remainder is: **8**
Subtract this power of 2 from the remainder and you get a new remainder: **2**

Therefore: **26 = 16 + 8 + 2 = 10000 + 1000 + 10 = 11010**

Alternate: 26 divided by two is 13, remainder 0. The ones digit is 0.
13 divided by two is 6, remainder 1. The twos digit is 1.
6 divided by two is 3, remainder 0. The fours digit is 0.
3 divided by two is 1, remainder 1. The eights digit is 1.
We reached the final 1. The sixteens digit is 1.
Therefore 26 in binary is 11010.

9-6. Binary Long Division

Long division works the same way in binary as in decimal notation. Binary notation can also be used for fractions.

a. (Read) In decimal notation, ½ is 0.5. What is one-half in binary?
You answer this question the same way you would in decimal. You can use long division.

$$\begin{array}{r} 0.1 \\ 10\overline{)1.0} \\ \underline{1.0} \\ 0.0 \end{array}$$

The result of long division is 0.1.

b. What is one-quarter in binary?

Solution:

$$\begin{array}{r} 0.01 \\ 100\overline{)1.00} \\ \underline{1.00} \\ 0 \end{array}$$

one-quarter is 0.01

c. What is three-quarters in binary? (Hint: add up the answers from (a) and (b)).

Solution: (using the hint) From part a, ½ is 0.1, and from part b, ¼ is 0.01. Therefore 3/4 = ½ + ¼ = 0.1 + 0.01 = 0.11

Alternate: (using long division):

$$\begin{array}{r} 0.11 \\ 100\overline{)11.00} \\ \underline{10.00} \\ 1.00 \\ \underline{1.00} \\ 0 \end{array}$$

(in binary ¾ is 11 divided by 100).

d. What is one-third in binary?

$$\begin{array}{r} 0.0101 \\ 11\overline{)1.0000} \\ \underline{0.1100} \\ 0.0100 \\ \underline{0.0011} \\ 0.0001 \end{array}$$

Solution: you can see that this is a repeating decimal. 1/3 is 0.010101...

If you know about geometric series, you can see that this repeating decimal corresponds to the series ¼ + $(¼)^2$ + $(¼)^3$ + ...

and the sum of this geometric series is: $\frac{1}{4}\left(\frac{1}{1-\frac{1}{4}}\right)=\frac{1}{4}\left(\frac{1}{\frac{3}{4}}\right)=\frac{1}{3}$

Refresher: A geometric series is a series of the form $1 + r + r^2 + r^3 + \ldots$ Where r is a number. If r is between 0 and 1, then the series adds up to a value S. The value of the geometric series is: S = 1/(1-r). The way you derive this formula is to observe that S – rS = 1, and solve for S: S(1 – r) = 1, and then S = 1/ (1 - r)

e. What is one-fifth in binary?

$$\begin{array}{r} 0.0011 \\ 101\overline{)1.0000000} \\ \underline{0.1010000} \\ 0.0110000 \\ \underline{0.0101000} \\ 0.0001000 \end{array}$$

Solution: this is a repeating decimal. 1/5 is 0.001100110011...

Extension: (very advanced for elementary school) Show that 0.0011... corresponds to: 3/16 [1 + 1/16 + $(1/16)^2$ + $(1/16)^3$ + ...]

9-7. Base Three

Base three notation uses the three Arabic numerals 0, 1, and 2. Using what you know from the previous work on binary, convert the first ten counting numbers to base three.

Number	Base 3 Sum	Base 3 Number
one	1	1
two	2	2
three	2 +1 10	10
four	10 2 +1 or +2	**11**
five	10 +2	**12**
six	12 +1	**20**
seven	11 +10 21	21
eight	21 +1 or ?	**22**
nine	22 +1 or ?	**100**
ten	12 +12 or ?	**101**

9-8. Divisibility in Base Three

These questions are about divisibility and division in base three.

a. What divisibility rule is there in base three for numbers that end with the digit 0? Write down a few base three numbers that end in 0 and spot the pattern.

Solution: The number three is 10 in base three and so multiples of three end in 0. In the table in problem 9-7, the ternary numbers 10, 20 and 100 are three, six, and nine.

b. How can you tell if a base three number is divisible by two (or even)? (Hint: the role of the number two in base three is similar to the role of the number nine in base ten—two is one less than three, just like nine is one less than ten. Think about the divisibility rule for nine.)

Solution: The decimal rule of ending in 0 or 2 does not apply in ternary numbers, because 10 is an odd number (three) but it ends in 0. The hint says to think of the number "2" like the decimal number "9". The divisibility rule for 9 in decimal is to add up the digits and see if the sum of the digits is a multiple of 9. This rule works for divisibility by two in ternary. Consider:

2 = 2 is even, sum of digits is 2.
4 = 11 is even, sum of digits in ternary is 2.
6 = 20 is even, sum of digits in ternary is 2.
8 = 22 is even, sum of digits in ternary is four, which is divisible by two.

Try a larger number: twenty-five. First convert to ternary using our dividing/remainder algorithm:
25 divided by three is 8, remainder 1. The ones digit is 1.
8 divided by three is 2, remainder 2. The threes digit is 2.
Our last digit is 2, smaller than three. The nines digit is 2.
In ternary, 25 is 221. The sum of the digits in ternary is five, odd, so 25 is odd.

c. What simple fraction is represented by the base-three number 0.1?

Solution: By analogy with other bases, the first decimal place is the one-thirds place. So 0.1 is the fraction 1/3.

d. In any notation system, some fractions are repeating fractions. In decimal, the fraction 1/3 is a repeating fraction, but the fraction 1/5 is not.

Write the fraction one-half in base three notation. (Use long division). Before you figure it out, do you expect it to repeat or not?

Solution: Since two and three are relatively prime (they have no factors in common except one) I expect one-half to repeat in base-three notation.

$$\begin{array}{r} 0.11 \\ 2\overline{)1.0000} \\ \underline{0.2000} \\ 0.1000 \\ \underline{0.0200} \\ 0.0100 \end{array}$$

This process repeats. ½ is 0.1111.... in base three.

Observation: The number two in ternary plays the role similar to 9 in decimal. The decimal expansion of 1/9 is 0.111... and the ternary expansion of ½ is 0.111...

9-9. Base Six Fractions

Base six notation uses the six Arabic numerals 0, 1, 2, 3, 4, and 5.

a. What is the fraction one-sixth in base six?

Solution: By analogy with other bases, one-sixth in base six would be 0.1.

b. What is the fraction one-half in base six?

Solution: One method would be to use long division. Another method is to recognize that one-half is three-sixths, and therefore in base six:
one-half = 0.1 + 0.1 + 0.1 = 0.3

Double checking by multiplication: In base six, 2 x 3 = six = 10.
Therefore 2 x 0.3 = 1.0

c. What is the fraction one-third in base six?

Solution: Using the same reasoning as in part b, one-third is two-sixths. Therefore one-third = 0.1 + 0.1 = 0.2 in base six.

d. Out of one-fourth and one-fifth, which do you expect to be a repeating fraction in base six? Why?

Solution: I expect one fifth to be a repeating fraction, because five is relatively prime to six. I do not expect one fourth to repeat, because four is a power of two, and two is a factor of six.

One fourth is nine thirty-sixths. Therefore, in base six:
one-fourth = 9/36 = 6/36 + 3/36 = 1/6 + 3/36 = 0.1 + 0.03 = 0.13

Double checking with long division:

$$\begin{array}{r} 0.13 \\ 4\overline{)1.000} \\ \underline{0.400} \\ 0.200 \\ \underline{0.200} \\ 0 \end{array}$$

e. The base six number 0.04 corresponds to what simple fraction in lowest terms?

Solution: In base six, 0.04 is four thirty-sixths, or 4/36. That corresponds to 1/9, one ninth.

9-10. Hexadecimal

The hexadecimal system is widely used in computer programming. It is base sixteen, and uses the sixteen digits 0, 1, 2, 3, 4, 5, 6, 7, 8, 9, A, B, C, D, E, and F.

a. What is the number six in hexadecimal? **Solution: 6**

b. What is the number twelve in hexadecimal? **Solution: C**

c. The hexadecimal number 13 is what number in decimal?

Solution: 13 in hexadecimal is sixteen + three = nineteen = 19 in decimal.

d. (Read) What is the fraction one-half in hexadecimal?

$$\begin{array}{r} 0.8 \\ 2\overline{)1.0} \\ \underline{1.0} \\ 0 \end{array}$$

Notice that in hexadecimal notation:

$$\begin{array}{r} 2 \\ \times\ 0.8 \\ \hline 1.0 \end{array}$$

Because 2 x 8 = sixteen = 10 in hexadecimal.

Solution: one half is 0.8 in hexadecimal.

e. What is the fraction one-sixteenth in hexadecimal?

Solution: By analogy with other bases, one sixteenth is 0.1 in hexadecimal (base sixteen).

f. What is the fraction one-eighth in hexadecimal? (To check your answer, look at the relationship between one eighth and one sixteenth.)

Solution: One eighth is two sixteenths. Therefore one eighth in hexadecimal is: 0.1 + 0.1 = 0.2

g. What is the fraction three-quarters in hexadecimal? ____________
(To check your answer: what is the relationship between three-quarters and one-sixteenth?)

Solution: Three quarters is twelve sixteenths. In hexadecimal, twelve is C. Therefore, three quarters is 0.C.

h. Think about what fractions repeat and which do not repeat in hexadecimal. What is one-third in hexadecimal?

Solution: The following numbers have factors that are not factors of sixteen: three, five, six, seven, nine, ten, eleven, twelve, thirteen, fourteen, and fifteen. I expect that the fractions one half, one quarter, one eighth, and one sixteenth would not repeat.

How To Deal With Base Number Problems

The following are only suggestions:

Learn the vocabulary:

- Base
- Binary
- Ternary
- Hexadecimal
- Divisibility
- Remainder
- Place value
- Power of a number

Look for patterns such as:

- In binary, even numbers end in zero.
- In ternary, multiples of three end in zero.
- If a number is written in base B, the place values to the **left** of the ones place are B, B^2, B^3, and so on.
- If a number is written in base B, the place values to the **right** of the ones place are $1/B$, $1/B^2$, $1/B^3$, and so on.

There might be surprises such as:

- In hexadecimal, one-half is 0.8, and 10 is sixteen. There are many surprises.
- The algorithm (recipes) for addition, subtraction, long multiplication, and long division are the same when applied to different bases.
- Divisibility rules are different when applied to different bases.

10. Polyhedra

Here are Professor Bear's reasons to for the tenth chapter to be about Polyhedra:

- Polyhedra are the next dimension up from polygons.
- Regular polyhedral have interesting properties. It's a curious mathematical fact that there are only five regular polyhedra.
- Understanding polyhedra requires different skills from numerical calculation. It helps to draw and build models and visualize.

Underpinnings of Polyhedra

To work with this section it's helpful to know how to plug numbers into a formula. It's also helpful to have experience with counting strategies. Some problems are related to Counting, some to Factors and Primes, and some to Area and Perimeter.

Outcomes

After doing this chapter students will be able to recognize some classic polyhedra: the five regular polyhedra and prisms. They'll have experience working with the geometrical ideas of length, area, volume, vertices, edges, faces, and polygons.

Polyhedra: Notes and Vocabulary

A **polyhedron** is a three-dimensional enclosed surface with flat faces and straight edges. The point where edges meet, a corner, is called a **vertex**.

A familiar example is the cube. Strictly speaking, the word polyhedron refers to the surface of the cube, not the inside of the cube.

Regular Polyhedra

A cube is a member of a special class of polyhedron called the Regular Polyhedra. The Regular Polyhedra have these properties:

- Every face is the same, and is a regular pentagon.

- The same number of edges meet at every vertex.
- The angle between nearest edges at each vertex is the same.

There are only five Regular Polyhedra:
Tetrahedron: four faces, each face is an equilateral triangle.

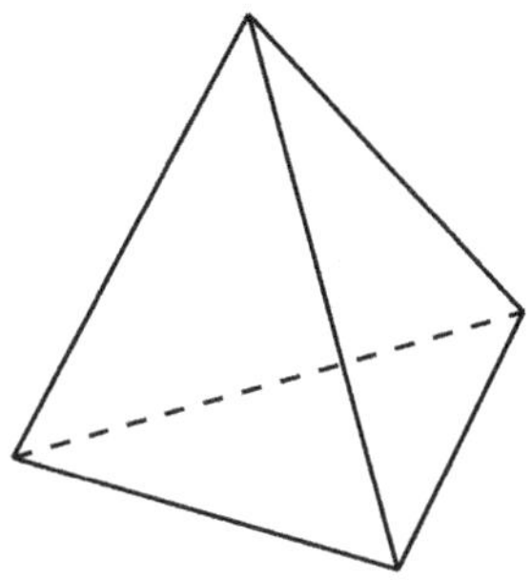

Cube: six faces, each face is a square. Octahedron: eight faces, each face is an equilateral triangle.

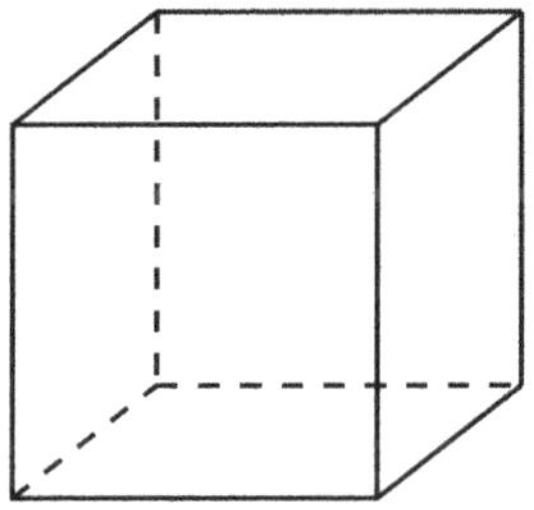

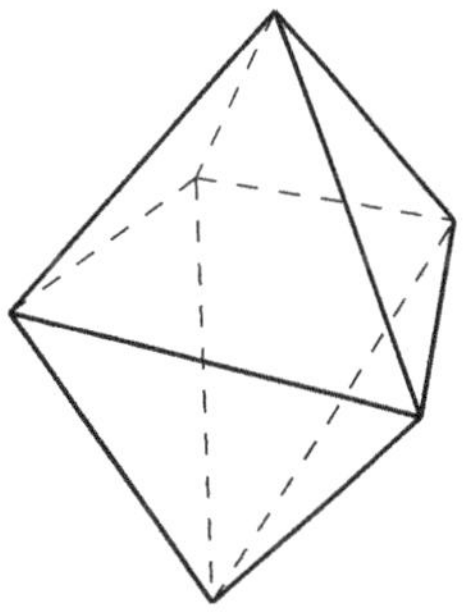

Dodecahedron: twelve faces, each face is a regular pentagon. Icosahedron: twenty faces, each face is an equilateral triangle.

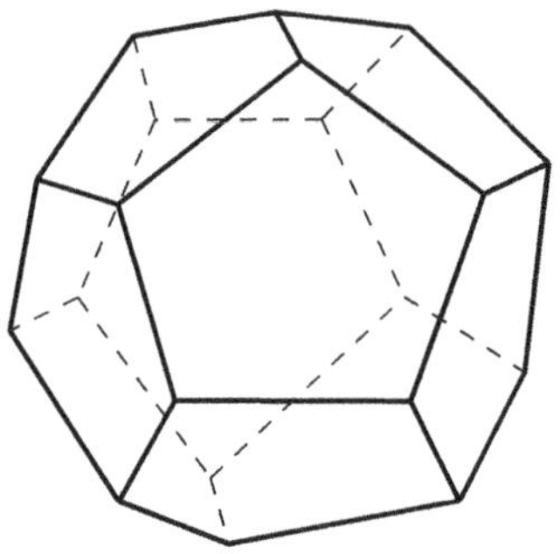

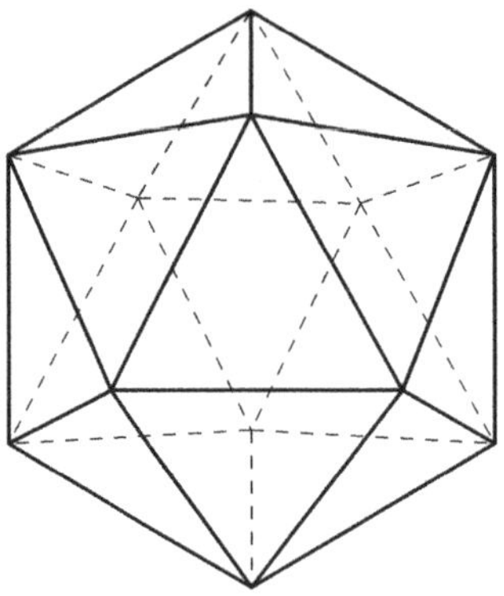

Prisms

A prism is a polyhedron that is similar to a cylinder. The difference between a prism and a cylinder is that if you slice a cylinder, you get a circle, but if you slice a prism, you get a polygon. Here are some examples of prisms.

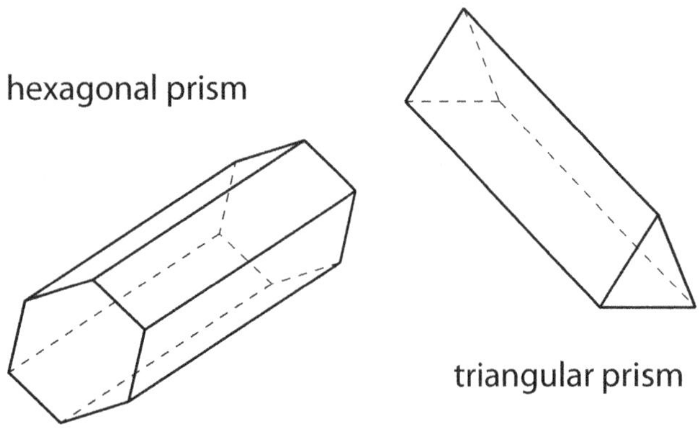

10-1. Cube Wrap

The following diagrams represent pieces of paper cut out along the outer lines. Which pieces of paper could be used to wrap a cube completely without any leftover paper, uncovered areas, overlapping, or tearing?

a.

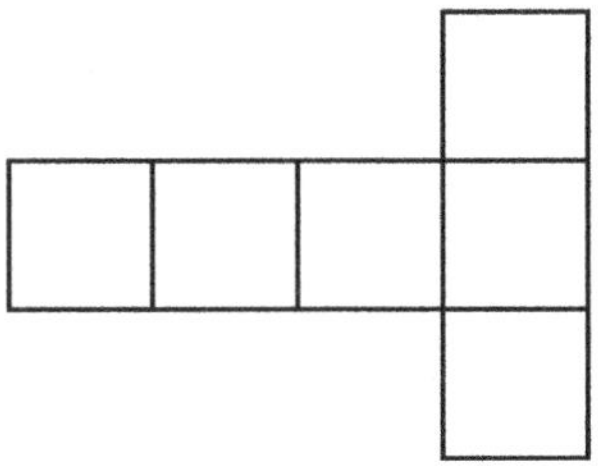

b.

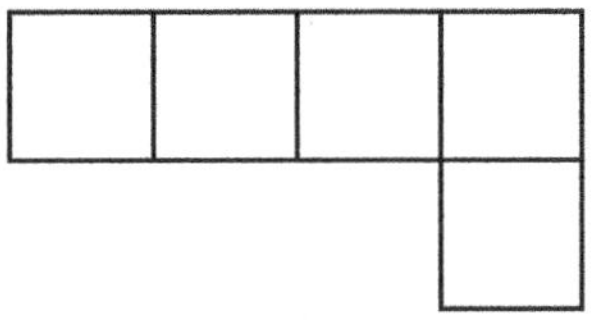

c.

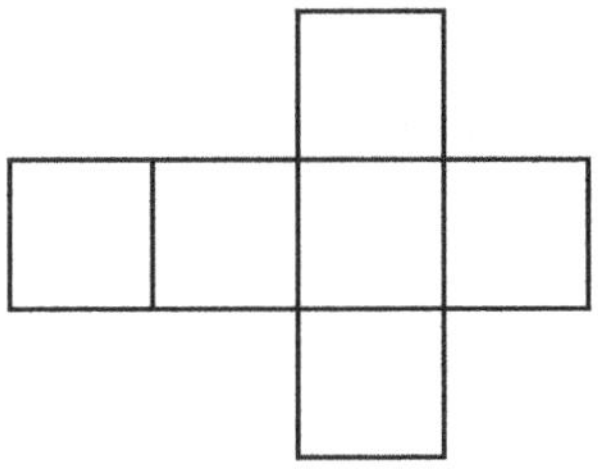

d.

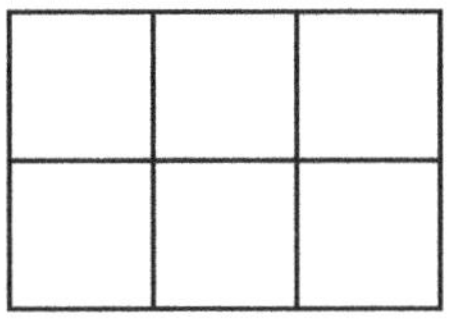

e.

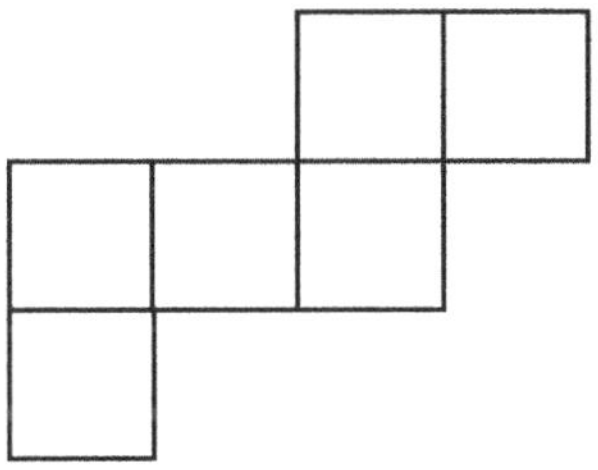

f.

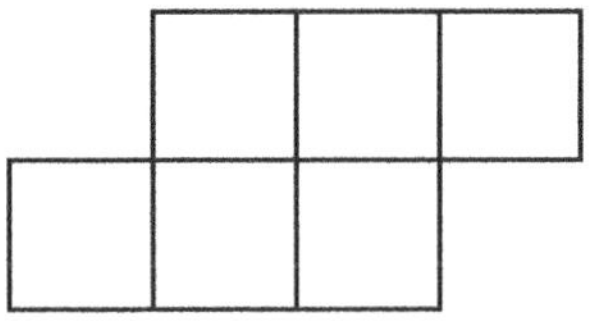

g.

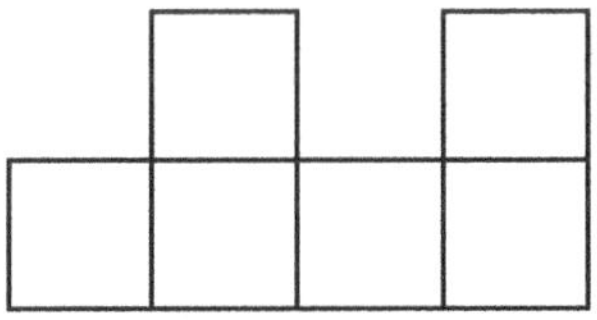

h.

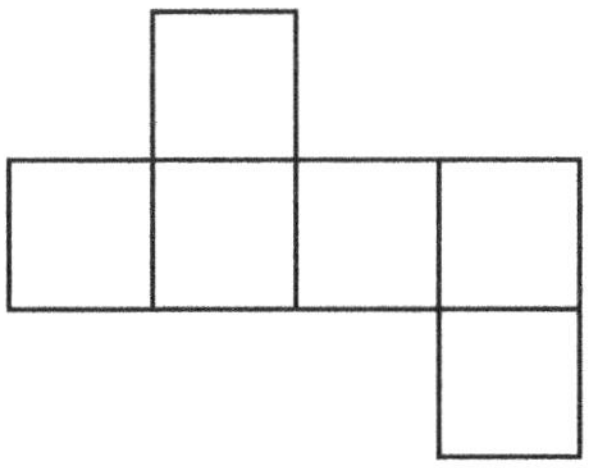

Solution: The pieces that would not wrap a cube are b, d, f, and g. The piece labeled b has the simple problem of not enough faces: it only has five. The best way to solve this problem is to attempt to wrap cubes with the pieces of paper, either with real paper or in your imagination. If you do not use real paper and cubes, you can attempt to label the faces and determine if a cube wrap is possible. For example, let B=bottom face, T=top face, and F1, F2, F3, and F4 be the side faces, in a clockwise direction.

a.

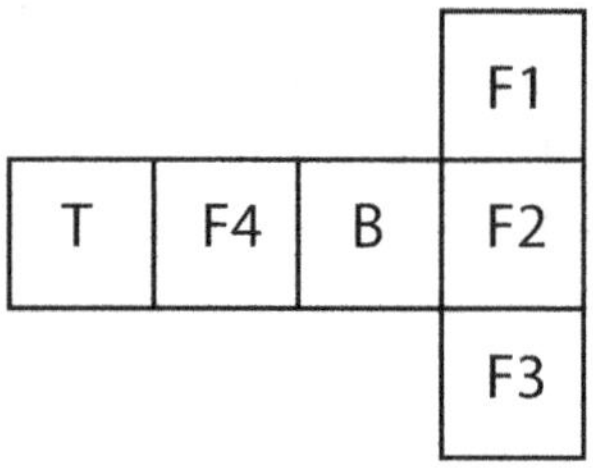

b.

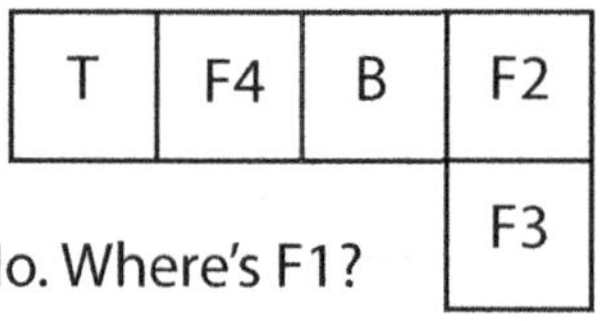

No. Where's F1?

c.

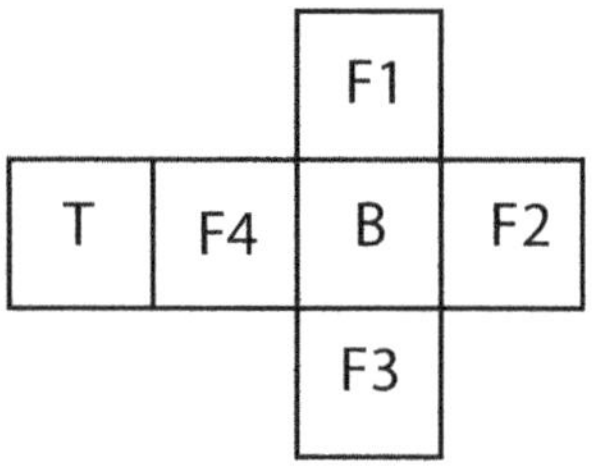

d.

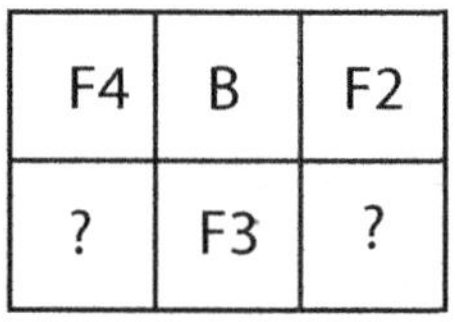

Does not work.

e.

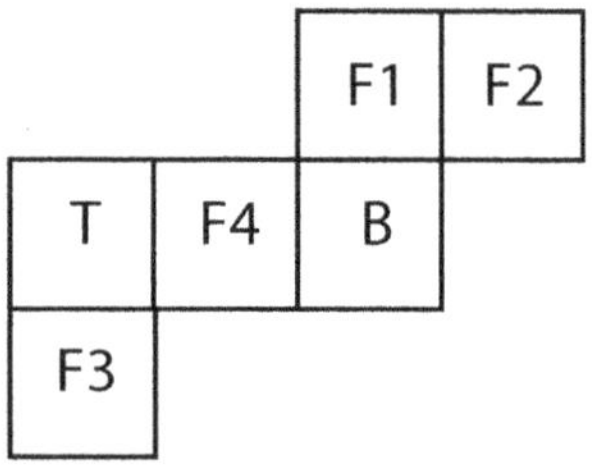

f.

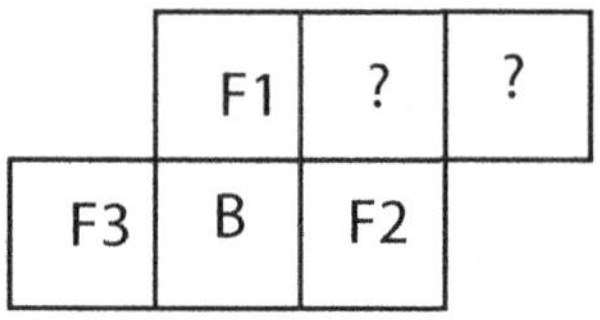

Does not work.

g.

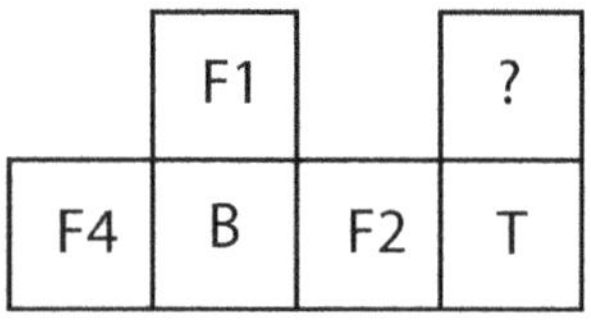

Does not work.

h.

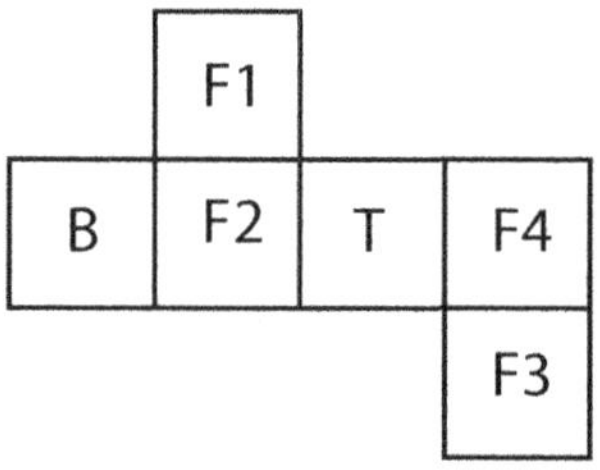

Note to Coach: There are multiple solutions for a, c, e, and h. The ones above were determined by choosing the location of "B" and then filling in the remaining faces and top. For b, d, f, and g it is necessary to show that no matter what choice is made for B, no solution is possible.

10-2. Face Paint

A large cube is made of little cubes as shown. It is four cubes wide, four cubes long, four cubes deep.

If the large cube is painted all over:

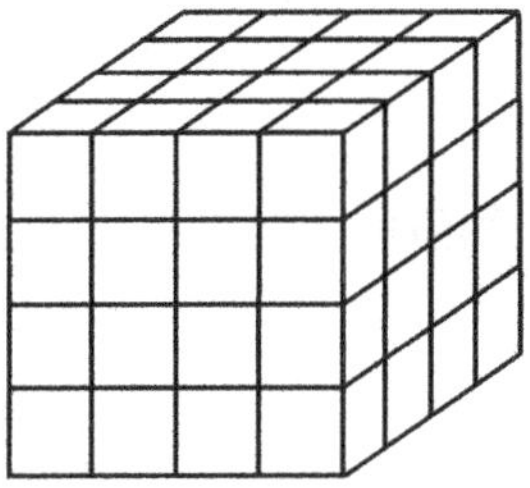

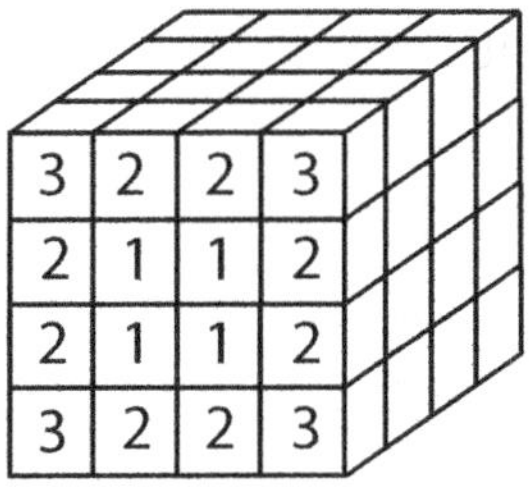

a. How many little cubes have paint on one face only?

Solution: The diagram above shows the number of faces painted on each little cube in the front face of the large cube. On each face of the big cube are four little cubes in the center with 1 painted face. There are six faces on the big cube. Therefore there are 4 x 6 = 24 little cubes with one painted face.

b. How many little cubes have paint on two faces only?

Solution: The diagram above shows that on each edge of the big cube, there are two little cubes with 2 painted faces. There are twelve edges on the big cube. Therefore there are 2 x 12 = 24 little cubes with two painted faces.

c. How many little cubes have paint on three faces only?

Solution: The diagram above shows that on each vertex (corner) of the big cube, there is one little cube with three painted faces. There are eight vertices on the big cube. Therefore there are 1 x 8 = 8 little cubes with three painted faces.

d. How many little cubes have no paint at all?

Solution: The little cubes with no paint cannot be seen in the diagram. They form a medium 2 x 2 x 2 cube within the big cube. The number of little cubes in the medium inner cube is 8.

To double check the solutions: the number of little cubes in the big cube is 4 x 4 x 4 = 64.

The total number of painted cubes is: 24 +24 + 8 = 56
The number of painted + unpainted cubes is 56 + 8 = 64

10-3. Three Views Of A Building

Rita uses some small cubes to build a structure. She takes photographs of her structure from three different views.

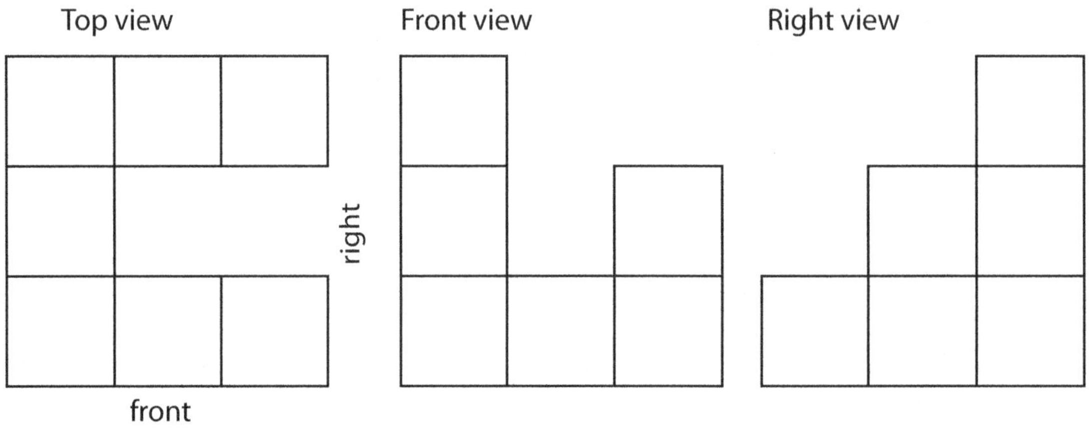

a. Label the top view with the height of each stack of cubes.

Solution: Because the right view shows the tallest stack of 3 cubes in the rear, and the front view shows the tallest stack on the left, the rear left stack must have a height of 3. The stack immediately to the right must have height 1, because the front view shows the middle height is 1. And the rear right must therefore have a height of 2, because the front view shows a height of 2 on the right. The right view shows a height of 1 on the entire front row, and the right view shows a height of 2 on the middle row.

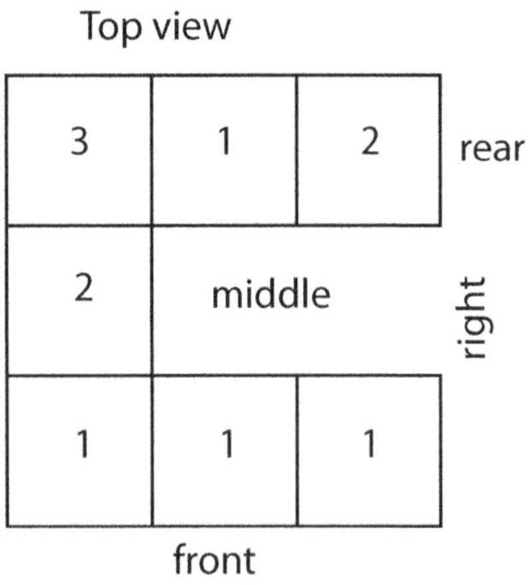

b. How many cubes did she use to build her structure?

Solution: From the diagram, she used 3 + 1 + 2 + 2 + 1 + 1 + 1 = 11 cubes.

10-4. Cube-O

Randall uses little cubes to build cube-Os.

How many cubes does it take to build a Twenty-O?

Solution: There are a number of approaches to this problem. If you have cubes for students to handle, they could simply build a Twenty-O and count. And that's not a bad thing. Professor Bear calls that the Brute Force method and it's very useful for students to have opportunities to see Brute Force in action and compare it with other methods.

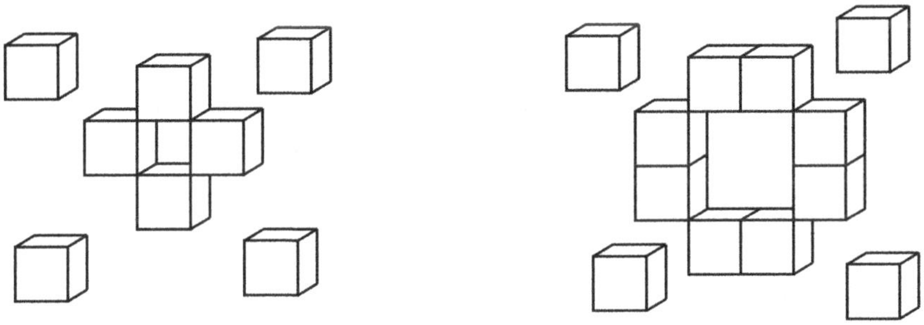

One-O has 4 + 4x1 cubes

Two-O has 4 + 4x2 cubes

Alternatively, the above diagram shows an approach for determining the number of cubes in an N-O. If a One-O has 4 + 4x1 cubes, and a Two-O has 4 + 4x2 cubes, you can see that a Three-O has 4 + 4x3 cubes, and so on. A Twenty-O has 4 + 4x20 cubes, which is 84 cubes.

Note to Coach: This problem is related to Sequences and Counting.

10-5. Tetrahedral Dice

Oliver has two tetrahedral dice, each marked 1, 2, 3, 4.

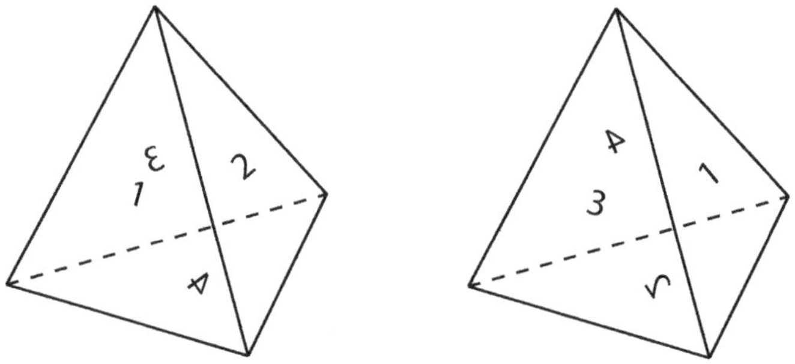

The dice land with one face on the table, and three faces showing.

a. What is the sample space of one roll of one tetrahedral die?

Solution: There is an immediate challenge here because the tetrahedral die does not have an up-most face. When it lands, one of the four faces is down, and three faces are up. It's great if you have tetrahedral models for students to look at. But if you don't, it's useful to draw several tetrahedra, with or without the dotted lines, and indicate what faces show and do not show. In the diagram above, the left die has 4 on the bottom, and shows 1-2-3. The right die has 2 on the bottom, and shows 3-1-4.

After students have a chance to explore these possibilities, the next challenge is how to label different rolls. The roll shown could be labeled "1-2-3" or "2-1-3" or "3-1-2". An easier way to label a roll uses the added total:
1-2-3 is represented by 6
1-2-4 is represented by 7
1-3-4 is represented by 8
2-3-4 is represented by 9

So there are 4 possible results for one die: 6, 7, 8, 9. This way of labeling has the advantage that it is shorter.

b. What is the sample space of one roll of two tetrahedral dice?

Solution: If you toss two tetrahedral dice, the sample space is (using the labeling method in part a):
6-6, 6-7, 6-8, 6-9, 7-6, 7-7, 7-8, 7-9,
8-6, 8-7, 8-8, 8-9, 9-6, 9-7, 9-8, 9-9.

Sixteen possibilities.

c. What is the probability of rolling a total of 16 in one roll of two tetrahedral dice? (16 would be the sum of all six exposed sides).

Solution: Looking at our sample space, there are three ways to get a total of 16: 7-9, 9-7, and 8-8. So the probability is 3/16.

d. What is the total that has the highest probability of occurring in one roll of two tetrahedral dice?

Solution: The totals, in our sample space, are:
12, 13, 14, 15, 13, 14 15, 16
14, 15, 16, 17, 15, 16, 17, 18

The total that shows up the most often is 15. It has probability ¼.

10-6. Cubes Made Out Of Cubes

Pamela has a set of identical wooden cubes one inch long on each side. She'd like to build a sequence of cubes, of side length 1, side length 2, side length 3, side length 4, and side length 5. How many cubes does she need to build them all?

Solution: The pattern here is that the cube of the side-length is needed to build the actual cube. So Pamela requires:

$$1^3 + 2^3 + 3^3 + 4^3 + 5^3 = 1 + 8 + 27 + 64 + 125 = 225$$

10-7. Parallelepipeds Out Of Cubes

Louise has 24 identical cubes, each with side length one inch.

a. How many different solid rectangular polyhedra (parallelepipeds) can she build with her cubes? (She must use all 24 cubes each time, but the different polyhedra can have different dimensions.)

Solution: The number 24 has a lot of factors. There are a set of height-one rectangular polyhedra she can make:
1x1x24, 1x2x12, 1x3x8, 1x4x6 (4)
Height-two polyhedra:
2x2x6, 2x3x4 (2)
A total of 6 different polyhedra can be constructed.

b. What is the volume of each polyhedron constructed in part a?

Solution: The volume of all the polyhedra is the same: 24 cubic inches.

c. What is the surface area of each polyhedron constructed in part a?

Solution: The surface areas are as follows:

1x1x24 has area 1 + 1 + 24 + 24 + 24 + 24 = 98 square inches.
1x2x12 has area 2 + 2 + 12 + 12 + 24 + 24 = 76 square inches.
1x3x8 has area 3 + 3 + 8 + 8 + 24 + 24 = 70 square inches.
1x4x6 has area 4 + 4 + 6 + 6 + 24 + 24 = 68 square inches.
2x2x6 has area 4 + 4 + 12 + 12 + 12 + 12 = 56 square inches.
2x3x4 has area 6 + 6 + 8 + 8 + 12 + 12 = 52 square inches.
It's great if the students illustrate their computations on diagrams.

Note to Coach: It's worth pointing out that once again this is an illustration of how the surface to volume ratio changes according to shape. Shape, unlike length or area, does not lend itself to quantitative measurement, and yet the surface:volume ratio is a number that tells something about the shape.

In this example, the surface to volume ratio changes from 98:24 (about 4.08) to 52:24 (about 2.17).

The surface to volume ratio is similar to the perimeter to area ratio.

10-8. Euler's Formula

A convex polyhedron is, roughly speaking, a polyhedron that, overall, bulges outward. It does not have any holes or caverns. The five Regular Polyhedra are all convex, and prisms are convex. The Swiss mathematician Leonhard Euler discovered a formula that is true for all convex Polyhedra:

$V - E + F = 2$

Where V is the number of vertices, E is the number of edges, and F is the number of faces. Show that Euler's formula is true for all of the Polyhedra below. Count the different numbers of vertices, edges and sides.

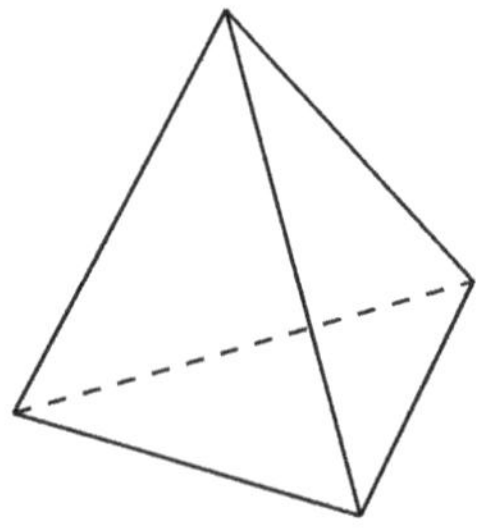

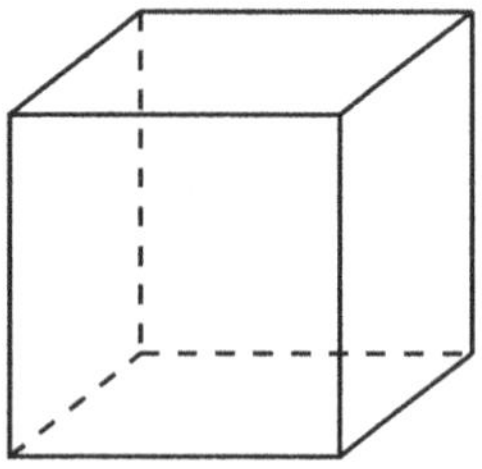

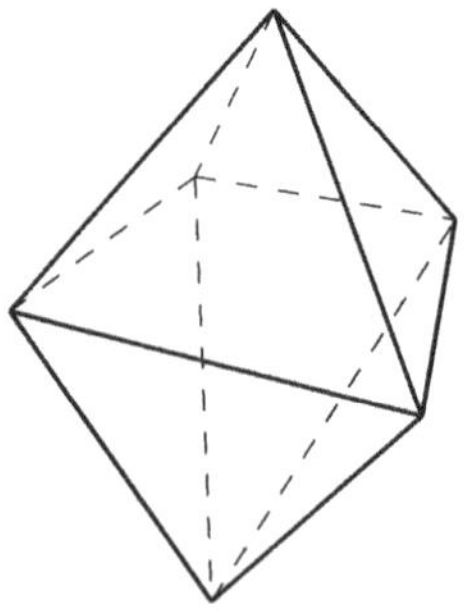

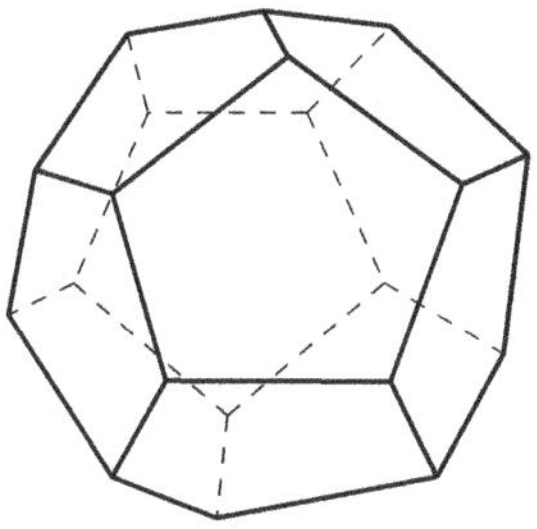

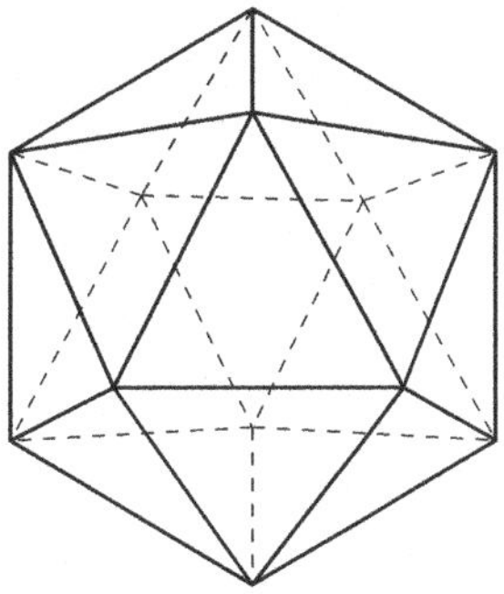

Solution: The details for each polyhedron are in this table:

Polyhedron	Vertices	Edges	Faces	V – E + F = ?
tetrahedron	4	6	4	4-6+4 = 2
cube	8	12	6	8-12+6 = 2
octahedron	6	12	8	6-12+8 = 2
dodecahedron	20	30	12	20-30+12= 2
icosahedron	12	30	20	12-30+20 = 2

Note to Coach: Proving Euler's Formula is quite far beyond this book (it can be done by mathematical induction on the number of vertices), but Euler's Formula can be used to prove an amazing fact: that only five regular polyhedra are possible. The proof is probably beyond what young students can do, but it is interesting to look at the argument:

Suppose you can build a regular polyhedron with F faces, E edges, and V vertices. Because the polyhedron is regular, every face is a regular n-gon (an n-gon is a polygon with n edges).

Every face is surrounded by n edges. But every edge is shared by two faces. Therefore, E, the number of edges, must be F x n x ½.

Every face is surrounded by n vertices. But because the polyhedron is regular, the same number of faces meets at every vertex. Let the number of faces that meet at each vertex be m.

Then V, the number of vertices, must be F x n x 1/m.

Euler's Formula states that V – E + F = 2, and therefore:

$\frac{Fn}{m}-\frac{Fn}{2}+F=2$ which factors into: $F\left(\frac{n}{m}-\frac{n}{2}+1\right)=2$

From this we can make a couple of observations:

Because F > 0, the expression $\frac{n}{m}-\frac{n}{2}+1$ has to be > 0.
The number n, the number of edges per face, must be at least 3. (There is no polygon with fewer sides than a triangle.)

This means: $\frac{n}{m}-\frac{n}{2}+1=n\left(\frac{1}{m}-\frac{1}{2}\right)+1>0$

With some algebra: $1>n\left(\frac{1}{2}-\frac{1}{m}\right)$ *and* $\frac{1}{n}>\left(\frac{1}{2}-\frac{1}{m}\right)$

Because $n\geq 3, \frac{1}{3}\geq\frac{1}{n}>\frac{1}{2}-\frac{1}{m}$ *which means* $\frac{1}{3}>\frac{1}{2}-\frac{1}{m}$

This in turn means:
$\frac{1}{m}>\frac{1}{2}-\frac{1}{3}=\frac{1}{6}$ and so 6 > m. The number m must be bigger than 2 (recall that m is the number of faces that meet at a vertex). So m can be 3, 4, or 5.

If m = 3, the formula $1>n\left(\frac{1}{2}-\frac{1}{m}\right)=n\left(\frac{1}{2}-\frac{1}{3}\right)=\frac{n}{6}$ means that n < 6.
Because n is at least 3, n can be 3, 4, or 5.

If m = 4, 1 > n/4 which means that n can only be 3.

If m = 5, 1 > 3n/10 which means that n can only be 3.

These five possibilities correspond to the five regular polyhedra. No others are possible.
Plugging into Euler's Formula:
If m = 3 and n =3, F = 4 (tetrahedron).
If m = 3 and n =4, F = 6 (cube)
If m = 3 and n = 5, F = 12 (dodecahedron)
If m = 4 and n = 3, F = 8 (octahedron)
If m = 5 and n = 3, F = 20 (icosahedron)

Note to Coach: There are two pairs of dual regular polyhedra. The midpoints of the faces of the cube form the vertices of an octahedron. The midpoints of the octahedral faces form the vertices of a cube. Similarly, the midpoints of the faces of the dodecahedron form the vertices of an icosahedron, and the midpoints of the icosahedron form the vertices of a dodecahedron. The tetrahedron is its own dual.

10-9. Octahedron Views

An octahedron is made up of two square pyramids joined together at the base:

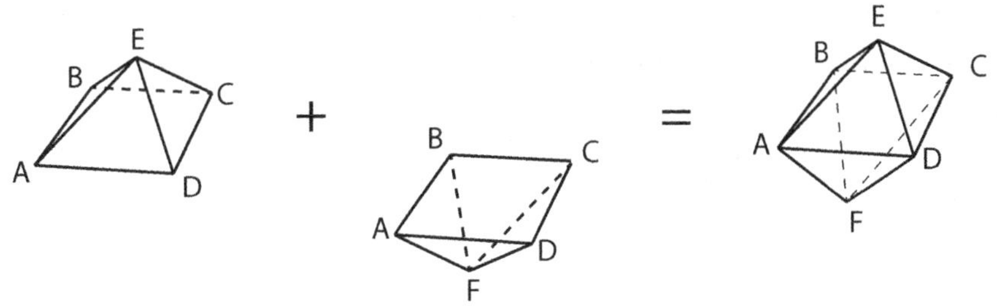

Here are some views of the octahedron from different directions. Label the vertices in these diagrams to match the labeling above.

Solution: There are a few different solutions for this problem. The following diagrams show appropriate solutions.

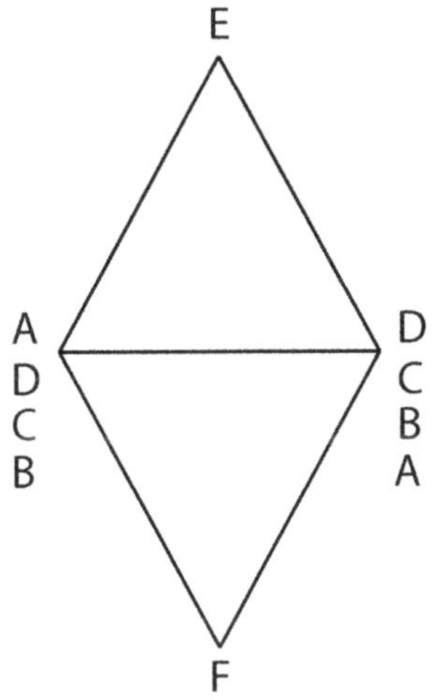

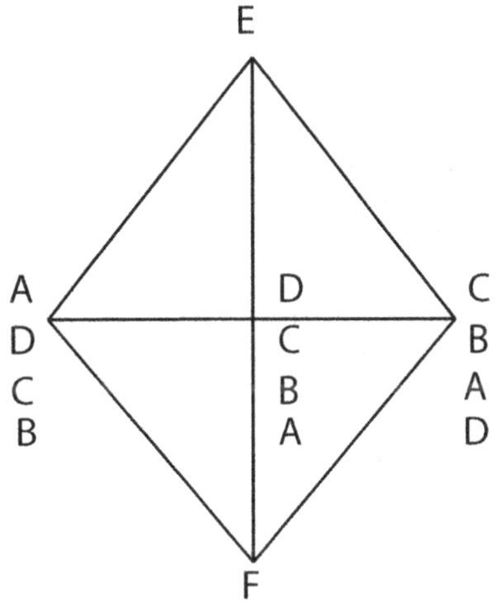

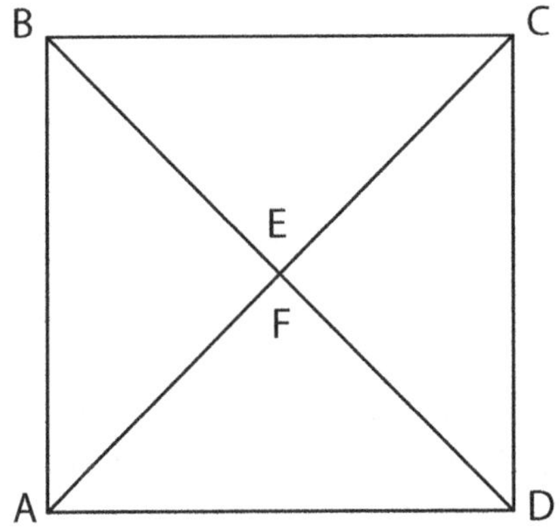

10-10. What Do You Get When You Slice A Cube?

Here are three different ways to slice a cube and get a new polyhedron. In each case, describe the resulting polyhedron. Count how many different polygons form the faces, and name the polygons. Evaluate Euler's formula for each sliced polyhedron.

a. Here is a cube with one corner removed.

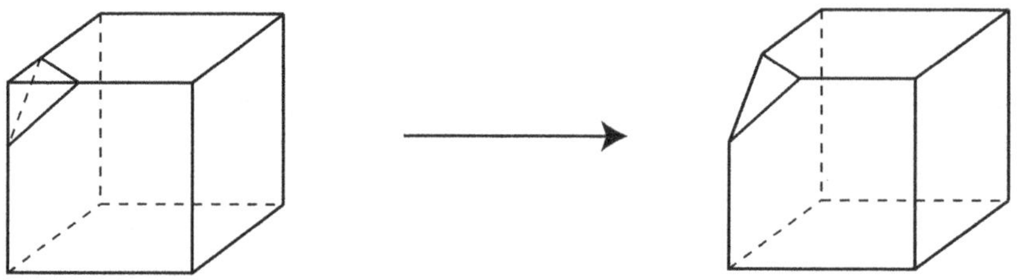

Solution: The original cube has 8 vertices, 12 edges, and 6 faces. After cutting, the polyhedron has 7 + 3 = 10 vertices, 15 edges, and 7 faces. Therefore Euler's Formula becomes: V – E + F = 10 – 15 + 7 = 2.

After cutting, the polyhedron's faces are: 3 irregular pentagons, 1 triangle, and 3 squares.

b. Here is a cube sliced in half along the top diagonal.

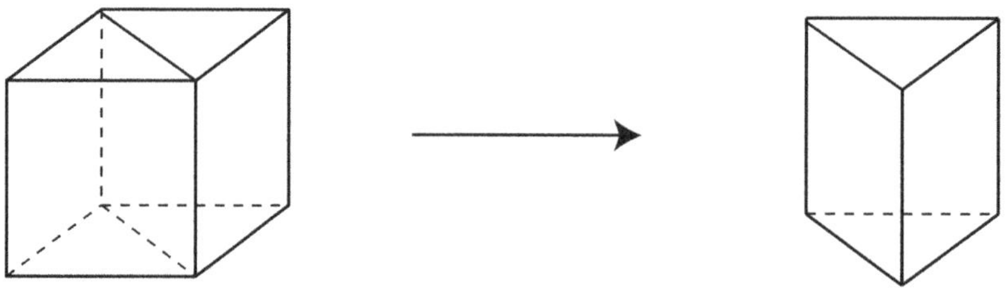

Solution: The original cube has 8 vertices, 12 edges, and 6 faces. After cutting, the polyhedron has 6 vertices, 9 edges, and 5 faces. Therefore Euler's Formula becomes: V – E + F = 6 – 9 + 5 = 2.

After cutting, the polyhedron's faces are: 1 rectangles, 2 squares, and 2 isosceles right triangles.

c. Here is a cube sliced at an angle.

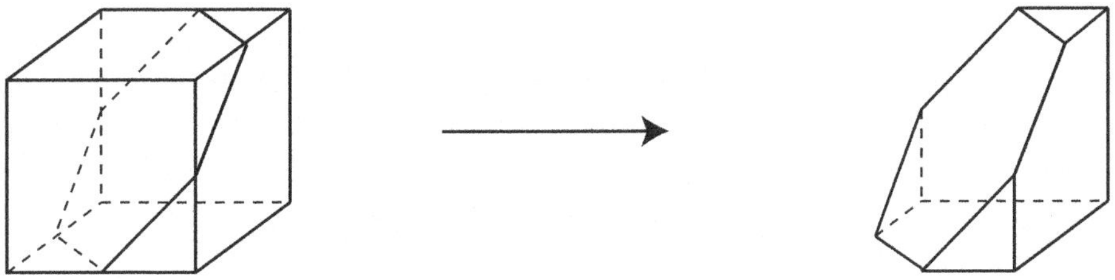

Solution: The original cube has 8 vertices, 12 edges, and 6 faces. After cutting, the polyhedron has 6 + 3 = 10 vertices, 15 edges, and 7 faces. Therefore Euler's Formula becomes: V – E + F = 10 – 15 + 7 = 2.

After cutting, the polyhedron's faces are: 1 hexagon, 3 right triangles, and 3 irregular pentagons.

Extensions: In this problem you can see that, with a straight cut through a cube, you can expose a triangle, rectangle, and hexagon. Are there any other polygons you can expose with a straight cut?

How To Deal With Polyhedra Problems

The following are only suggestions:

Learn the vocabulary:

- Polygon
- Polyhedron
- Face
- Edge
- Vertex
- Cube
- Tetrahedron
- Octahedron
- Dodecahedron
- Icosahedron
- Parallelepiped

Look for patterns such as:

- Volumes of solid cubes are cubic numbers 1, 8, 27, 64, 125, ...

- The closer a polyhedron is to a cube, the smaller its surface to volume ratio; in other words, the closer the shape is to a cube, the smaller area it has for a given volume.

There might be surprises such as:

- From one direction, an octahedron looks like a square.
- From one direction, an icosahedron looks like a hexagon.
- It's possible to slice a cube into a prism.
- It's possible to slice a cube and expose a hexagon.

References

Arnold, B. [1962]. *Intuitive Concepts in Elementary Topology*. Prentice-Hall, Inc.

Gelfand, I. M., and Shen, A. [2004]. *Algebra*. Birkhäuser Boston, Inc.

Lencher, G. [2005]. *Creative Problem Solving in School Mathematics, Second Edition.* Mathematical Olympiads For Elementary and Middle Schools, Inc.

Index

Made in the USA
Monee, IL
05 November 2026

67377204R10079